T0258576

Advances
in Network
Management

Advances in Network Management

Jianguo Ding

CRC Press
Taylor & Francis Group
Boca Raton London New York

CRC Press is an imprint of the
Taylor & Francis Group, an **informa** business
AN AUERBACH BOOK

Auerbach Publications
Taylor & Francis Group
6000 Broken Sound Parkway NW, Suite 300
Boca Raton, FL 33487-2742

© 2010 by Taylor and Francis Group, LLC
Auerbach Publications is an imprint of Taylor & Francis Group, an Informa business

No claim to original U.S. Government works

Printed in the United States of America on acid-free paper
10 9 8 7 6 5 4 3 2 1

International Standard Book Number: 978-1-4200-6452-0 (Hardback)

Visit the Taylor & Francis Web site at
http://www.taylorandfrancis.com

and the Auerbach Web site at
http://www.auerbach-publications.com

Contents

List of Figures

List of Tables

Foreword I

Like computer security, network management is a latent issue that computer and network users tend to ignore—until a disruption occurs. But today's business applications depend on reliable, secure, and well-performing networked computer infrastructures that may span large geographical areas and a multitude of management domains. Such infrastructures constantly change because hardware and software components are replaced or updated, new components are added, and old ones are shut down. In addition, hardware is subject to degradation and failure, software components give rise to faults, and the network may exhibit performance bottlenecks. It is obvious that successful network management plays a crucial economic role.

System administrators have a variety of hardware and software tools available to monitor the resources in their management domain and to help in diagnosing and reacting to defects—often after the fact. However, it would not be much more effective when potential hardware and software malfunctions, failures, and other threads could be foreseen and protective measures could be taken before they can occur? The growing complexity of today's distributed computing infrastructures requires innovative predictive techniques and self-managing mechanisms.

Although network management came to life in AT&T's telecommunications network already in 1920, this book is timely in addressing a contemporary view on network management. The book provides insight into fundamental concepts of network management. It presents a range of theories and practical techniques for network management and discusses advanced networks and network services. The final chapter covers advanced paradigms such as autonomic computing, context-aware systems management and automatic techniques aiming at self-management, self-(re)configuration, self-optimization, self-healing, or self-protection.

Autonomic computing deals with the increasing difficulty of managing distributed computing systems, which become more and more interconnected and diverse, and software applications that go beyond corporate boundaries into the Internet. Autonomic computing anticipates systems consisting of myriads of interacting autonomous components that are able to manage themselves according to predefined goals and policies.

Context-aware systems management aims to take the availability of dynamically changing resources and services during a particular period of system operation into account. Management policies and automated mechanisms need to be able to adapt to dynamic changes such that they offer the best possible level of service to the user in the given situation.

Self-management capabilities include automated configuration of components, automated recognition of opportunities for performance improvement (self-optimization), automatic detection, diagnosis and repair of local hardware and software problems, and automatic defense against attacks. The objective is to minimize the degree of external intervention, e.g., by human system managers, and, at the same time, to preserve the architectural properties imposed by its specification. The concept of self-configuration refers to a process in which an application's internal structure changes depending on its environment.

With its breadth and depth in theoretical, technical, and research topics, the book serves well as an account of both the state-of-technique and state-of-the-art in a very important and current research area.

Prof. Dr.-Ing. Bernd J. Krämer
FernUniversität in Hagen, Germany

April 2009

Foreword II

The Challenges of IT Investments: Management and Sustainability

It is very challenging to find stats that clearly spell out the bigger picture of IT investments and the impact on their management and maintenance. The only relevant stats that summoned the issue very well was one from Jawad Khaki, VP at Microsoft responsible for networking where he shows that the US business volume of hardware and software has massively shrunk over the past two decades from $200 billion down to $60 billion, while the cost of management and support has exploded from less that $20 billion to $140 billion. 70% of it is spent on the management and maintenance of the legacy networks and only 30% is spent on managing new networks.

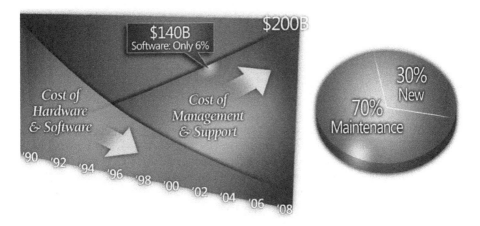

Figure 1: IT Complexity and Cost

The only other sector that has a similar picture is the airlines sector, although there are less planes crashing daily than servers and computer systems. The airlines sector has the stringent safety and security standards to follow, making air transportation the safest sector, at least compared to the automobile sector. However, if the same safety and security standards were applied to the IT sector, then either the investment in management and maintenance would be dramatically increased and make the computer sector inefficient or the IT sector would need to re-think its way forward on how to limit the explosion of its sustainability costs. The deeper impact of such a picture has significant strategic implications on the IT sector. The CTOs are absorbed more in day-to-day functions and lose the strategic thinking about introducing new technologies and new efficiencies.

The outsourcing of tedious tasks have resolved some issues but introduced new ones as well, such as making the introduction of new technologies very difficult because the outsourcer would tag the new technologies at a higher price.

The discussions among the researchers to simplify management or even to enable self-management or discuss implementation of autonomicity in networks have been one of the favorite research topics because the benefits are quite evident. The commoditization of networking going into every sector and even to the home is racing in front of us.

The Internet has reached a high level of penetration, between 50% and 75% in Europe and in the US. This makes the Internet the next technology to become a utility with over 1.5 billion users around the world. Cell phones are used now by over 4.0 billion users, although 10% of them are smart phones using Internet services. This level of penetration will put a lot more pressure on the management and maintenance on the networking side.

Network management is a domain plagued by proprietary solutions and some oligopolistic solutions. The industry should strive to create open source network management solutions that allow for easier integration of all vendors and applications developers. Another approach to use in parallel is to define generic autonomic network architectures (GANA) such as those defined by the Autonomic Future Internet (AFI) Industry Specification Group under the auspices of ETSI (the European Telecommunications Standards Institute) coordinated by the EU-funded project EFIPSANS (Exposing the Features in IP version Six protocols that can be exploited /extended for the purposes of designing / building Autonomic Networks and Services). AFI is working on the GANA vision to enable self-management concepts that will allow for the next step in self-managed networks, a promising path for this complex and legacy-plagued networking industry. AFI is now calling for contributions to the GANA specifications, and the definition of an evolutionary path toward self-managing future multi-services (i.e., Future Internet), from Future Internet research architects across the globe.

One of the new key technologies that needs close attention is the new Internet Protocol version 6 (IPv6). The deployment of IPv6 has become an issue of strategic importance for many economies. Enterprise networks and ISPs play a key role in ensuring the availability of this new protocol on their networks. It cannot be denied that the complexities exist in deploying IPv6 in an IPv4 world. Knowing this, telecom operators and ISPs have to ensure a viable transition strategy that takes into account transparent interoperability and mature integrated functionalities for deploying advanced applications on both IPv4 and IPv6. This potent combination will enable operators and ISPs to exploit the richer services offered by IPv6 while interoperating with IPv4 during this long transition period, creating new business models that will generate return on investment without waiting for the whole world to be fully IPv6 deployed. The management using IPv6 will be a lot more transparent to the network administrators as the reachability of any node on the network is an essential element in his work to check on each device, to update its security, and to manage it from any remote place.

The author raises essential issues at stake and brings the necessary expertise and experience identifying the challenges and proposing recommendations of great value to a world made of heterogeneous and widely uninteroperable networks designed with private addressing schemes that inhibit end-to-end management and end-to-end services.

Latif Ladid
President, IPv6 Forum
September 2009

Preface

Network management is facing new challenges, stemming from the growth in size, heterogeneity, pervasiveness, complexity of applications, network services, the combination of rapidly evolving technologies, and increased requirements from corporate customers.

Over a decade ago, the classic agent-manager centralized paradigm was the prevalent network management architecture, exemplified in the OSI reference model, the Simple Network Management Protocol (SNMP), and the Telecommunications Management Network (TMN) management framework. The increasing trend toward enterprise application integration based on loosely coupled heterogeneous IT infrastructures forces a change in management paradigms from centralized and local to distributed management strategies and solutions that are able to cope with multiple autonomous management domains and possibly conflicting management policies. In addition, service-oriented business applications come with end-to-end application-level quality of service (QoS) requirements and service-level agreements (SLA) that depend on the qualities of the underlying IT infrastructure.

More recently, requirements in network management and control have been amended by emerging network and computing models, including wireless networks, ad hoc networks, overlay networks, Grid networks, optical networks, multimedia networks, storage networks, the convergence of next generation networks (NGN), or even nanonetworks, etc. Increasingly ubiquitous network environments require new management strategies that can cope with resource constraints, multi-federated operations, scalability, dependability, context awareness, security, mobility, and probability, etc.

A set of enabling technologies are recognized to be potential candidates for distributed network management, such as policy-based management strategies, artificial intelligence techniques, probabilistic approaches, web-based techniques, agent techniques, distributed object computing technology, active networks technology, bio-inspired approach, or economic theory, etc. To bring complex network systems under control, it is necessary for the IT industry to move to autonomic management, context-aware management, and self-management systems in which technology itself is used to manage technology.

This book documents the evolution of networks and the trends, the evolution of network management solutions in network management paradigms, protocols, and techniques. It also investigates novel management strategies for emerging networks. The areas covered range from basic concepts to research-level material, including future directions.

The targeted audience for this book includes researchers (faculty members, PhD students, graduate students, senior undergraduates); professionals who are working in the area of network management; engineers who are designers or planners for network management systems; and those who would like to learn about this field.

About the Author

Jianguo Ding holds the degree of a Doctor Engineer (Dr.-Ing.) in Electronic Engineering from the Faculty of Mathematics and Computer Science of FernUniversität in Hagen, Germany.

After concluding his PhD research he was awarded the ERCIM (European Research Consortium for Informatics and Mathematics) "Alain Bensoussan" postdoctoral fellowship, which supported his research work at the Faculty of Science, Technology and Communication (FSTC) at the University of Luxembourg between July 2008 and April 2009. From May 2009 until January 2010 he continues to work as an ERCIM postdoctoral research fellow at the Department of Electronics and Telecommunications of the Norwegian University of Science and Technology (NTNU).

From 2005 to July 2008, he was a lecturer and an associate professor at the Software Engineering Institute of East China Normal University, China.

From November 2005 to February 2006 he was awarded the International research grant from University of Genoa in Italy and worked at Department of Communication, Computer and System Science (DIST), University of Genoa for the cooperation research in wireless and mobile networks.

From August 2001 to March 2004, he was awarded a DAAD (the German Academic Exchange Service) doctoral scholarship and studied at Department of Electrical and Computer Engineering of FernUniversität in Hagen, Germany.

Jianguo Ding is a Member of the IEEE and a Member of the ACM. His current research interests include network management and control, wireless and mobile networks, network security, network performance evaluation, intelligent technology, and probabilistic reasoning. Dr. Ding has published several book chapters, journal papers and peer-reviewed conference papers in these areas.

Acknowledgments

It has been a pleasure to work with Dr. Richard O'Hanley and Ms. Stephanie Morkert of Taylor & Francis Group. I want to thank them for their professional support and encouragement in publishing this book.

I would like to acknowledge Prof. Bernd J. Krämer for his help with the book proposal and the valuable advice with the book content.

Many thanks are due to Prof. Pascal Bouvry, Prof. Franco Davoli, Prof. Ilangko Balasingham, and Ranganai Chaparadza for their great support and help in preparing the book.

I owe a great deal of thanks to all my colleagues at the University of Luxembourg and the Norwegian University of Science and Technology for their kind help and insightful discussions.

I must not fail to mention those researchers and publishers who supported me by providing original materials and approving copyright permission, including the IEEE, the ACM, Wiley & Sons, Springer, Elsevier, the ETRI Journal, and Mr. Robert H Zakon. I would like to express my gratitude to the anonymous authors of formal or informal white papers, technical reports, and web contents that are used in my book.

This work was supported by ERCIM (the European Research Consortium for Informatics and Mathematics), EU FP7 project EFIPSANS (Exposing the Features in IP version Six protocols), UL project EVOSEC F1R-CRC-PUL-08EVOS, NSFC 60773093, Natural Science Foundation 08ZR1407200 and project 973 2005CB321904.

Finally, I appreciate my family for their great patience and enormous love throughout the book publishing work. The book writing accompanied Zirui Ding's birth and growth.

Jianguo Ding

Jianguo.Ding@ieee.org

September 2009

Curiosity, or love of the knowledge of causes, draws a man from consideration of the effect to seek the cause; and again, the cause of that cause; till of necessity he must come to this thought at last, that there is some cause whereof there is no former cause, but is eternal; which is it men call God. So that it is impossible to make any profound inquiry into natural causes without being inclined to believe there is one God eternal.

Thomas Hobbes (1588 – 1679)

LEVIATHAN

Chapter XI: Of the Difference of Manners

PART I: Of Man

Chapter 1

Introduction

This book aims to provide a wide coverage of key technologies in network management including network management architectures and protocols, theories and techniques, the management of emerging networks, and autonomic and self-management.

1.1 Motivation of the Book

Current networks are evolving rapidly and have shown many new characteristics and are expected to support multiple emerging services. A variety of challenges in networks mean current management approaches have to improve to match the new challenges and requirement.

The techniques update so quickly that lots of new concepts and techniques bring confusions and challenges to an audience, even for professionals.

This book provides the reader the complete overview of evolutions in networks and network management and a deeper insight into network management. It helps those who seek challenging topics that define or extend frontiers of the technology.

After the study, the reader can get clear ideas of network history, network evolution and the trends, the emerged network management requirement and the available theory, and techniques and strategies for network management systematically.

Further, it enables the reader to clarify the basic concepts and theories in the area of network management, and to understand which and how management challenges can be resolved by appropriated methods. It also guides the reader to trade-off and select potential approaches for future management challenges, which come from ongoing evolution of networks.

1.2 Structure and Organization of the Book

This book is organized based on the order of knowledge logic in networks and network management. It allows the reader to study step by step and understand the reasonable relationship between challenges and resolutions in network management.

Lots of basic concepts that emerged in networks and network management are clearly defined. Clear classification and much comparison allow the reader to catch the main ideas of every theory and technology easily.

The whole book is organized as follows;

Chapter 2 presents the evolution of networks, a detailed survey of tele-communication networks, computer networks, network architectures, and the future on networks.

Chapter 3 illustrates the evolution of network management, including network management architectures, network management protocols, and the network management functions.

Chapter 4 investigates the theories and techniques for network manage-ment, including policy-based approach, artificial intelligence techniques, graph-theoretic techniques, probabilistic models, web-based approach, agent techniques, distributed object computing techniques, active network tech-nique, bi-inspired approach, XML-based approach, etc.

Chapter 5 investigates the management strategies for emerging networks and services, such as wireless and ad hoc networks, overlay networks, opti-cal network, grid networks, storage networks, multimedia networks, satellite networks, cognitive networks, and future Internet.

Chapter 6 presents topics in autonomic computing and self-management, which might be efficient strategies in managing increasing complex networks. These topics include autonomic management, context-aware management, self-management, and automatic management.

Chapter 2

Evolution of Networks

2.1 Introduction of Networks

2.1.1 Definition of Networks

In general, the term network can refer to any interconnected group or system. More specifically, a network is any method of sharing information between two systems. In information technology, a network is a series of points or nodes interconnected by communication paths. Networks can interconnect with other networks and contain subnetworks. Digital networks may consist of one or more routers that route data to the correct user. An analogue network may consist of one or more switches that establish a connection between two or more users. For both types of network, a repeater may be necessary to amplify or recreate the signal when it is being transmitted over long distances. This is to combat attenuation that can render the signal indistinguishable from noise [ATIS01].

Often the term networks and communication networks are used interchangeably. In current information area, networks are identified as telecommunications networks and computer networks. Sometimes, networks can be classified based on the communication technologies that are used to build the networks, see Figure 2.1[Per05]. In this book, if no specific designation, the term networks denote both telecommunication networks and computer networks.

Telecommunications Network

A telecommunications network is a network of telecommunications that messages may be passed from one part of the network to another over multiple links and through various nodes. The process of information exchange among the telecommunications network typically involves the sending of electromagnetic waves by electronic transmitters, but in earlier years it may have involved

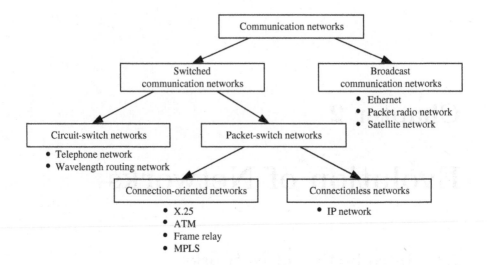

Figure 2.1: A Classification of Communication Networks

the use of smoke signals, drums, or semaphore. Today, telecommunication is widespread and devices that assist the process, such as the television, radio, and telephone, are common in many parts of the world. There are also many networks that connect these devices, including computer networks, public telephone networks, radio networks, and television networks. Computer communication across the Internet is one of many examples of telecommunication.

Early inventors of telecommunication systems include Alexander Bell, Guglielmo Marconi, and John Logie Baird. Typical telecommunication example is telephone networks.

Computer Networks

A computer network is a collection of computer systems or devices connected to each other. The computer network allows computers to communicate with each other and share resources and information.

A computer network generally involves at least two devices capable of being networked with at least one usually being a computer. The devices can be separated by a few meters (e.g., via Bluetooth) or nearly unlimited distances (e.g., via the Internet). Computer networking is sometimes considered a sub-discipline of telecommunications, and sometimes of computer science, information technology, and computer engineering. Computer networks rely heavily upon the theoretical and practical application of scientific and engineering disciplines.

Examples of networks are the Internet, a wide area network that is the largest to ever exist, or a small home local area network (LAN) with two computers connected with standard networking cables connecting to a network interface card in each computer, or a sensor network, which consists of spatially distributed autonomous devices using sensors to cooperatively monitor physical or environmental conditions, such as temperature, sound, vibration, pressure, motion or pollutants, at different locations, etc.

2.1.2 Network Topologies and Functions

Network Topologies

A topology is basically a map of a network. There are three basic categories of network topologies: physical topologies, signal topologies, and logical topologies.

Physical topology describes the layout of the cables and workstations and the location of all network components. Often, physical topologies are compared to logical topologies, which define how the information or data flows within the network.

Signal topology describes the mapping of the actual connections between the nodes of a network, as evidenced by the path that the signals take when propagating between the nodes.

Logical topology describes the mapping of the apparent connections between the nodes of a network, as evidenced by the path that data appear to take when traveling between the nodes.

The logical classification of network topologies generally follows the same classifications as those in the physical classifications of network topologies, the path that the data take between nodes being used to determine the topology as opposed to the actual physical connections being used to determine the topology.

- Logical topologies are often closely associated with media access control (MAC) methods and protocols.

- The logical topologies are generally determined by network protocols as opposed to being determined by the physical layout of cables, wires, and network devices or by the flow of the electrical signals, although in many cases the paths that the electrical signals take between nodes may closely match the logical flow of data, hence the convention of using the terms "logical topology" and "signal topology" interchangeably.

- Logical topologies are able to be dynamically reconfigured by special types of equipment such as routers and switches.

It is important to note that a network can have one type of physical topology and a completely different logical topology [GS05]. Two networks have the same topology if the connection configuration is the same, although the networks may differ in physical interconnections, distances between nodes, transmission rates, and/or signal types.

The common types of network topology are illustrated and defined as below:

Bus topology: A network topology in which all nodes, i.e., stations, are connected together by a single bus. Bus topology is sometimes called linear topology. See Figure 2.2.

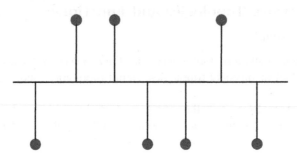

Figure 2.2: Bus Topology

Star topology: A network topology in which peripheral nodes are connected to a central node, which rebroadcasts all transmissions received from any peripheral node to all peripheral nodes on the network, including the originating node.

All peripheral nodes may thus communicate with all others by transmitting to, and receiving from, the central node only.

The failure of a transmission line, i.e., channel, linking any peripheral node to the central node will result in the isolation of that peripheral node from all others.

If the star central node is passive, the originating node must be able to tolerate the reception of an echo of its own transmission, delayed by the two-way transmission time, i.e., to and from the central node, plus any delay generated in the central node. An active star network has an active central node that usually has the means to prevent echo-related problems. See Figure 2.3.

Tree topology: Also known as hierarchical, the type of network topology in which a central "root" node (the top level of the hierarchy) is connected to one or more other nodes that are one level lower in the hierarchy (i.e., the second level) with a point-to-point link between each of the second level nodes and the top-level central "root" node, while each of the second level nodes that are connected to the top-level central "root" node will also have one or more other nodes that are one-level lower in the hierarchy (i.e., the third level) connected to it, also with a point-to-point link, the top-level central "root" node being the only node that has no other node above it in the hierarchy.

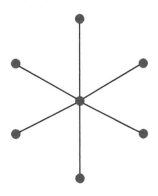

Figure 2.3: Star Topology

See Figure 2.4.

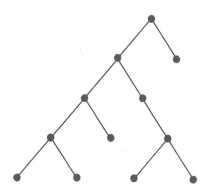

Figure 2.4: Tree Topology

Ring topology: A network topology in which every node has exactly two branches connected to it. Ring topology is classified as **Single ring topology** and **Dual ring topology**. Dual ring topology is often used in optical networks because of its self-healing feature. See Figure 2.5.

Fully connected topology: A network topology in which there is a direct path between any two nodes. In a fully connected network with n nodes, there are $\frac{n(n-1)}{2}$ direct paths in the whole network. A network with fully connected topology is sometimes called fully connected mesh network. See Figure 2.6.

Hybrid topology: A combination of any two or more network topologies. Instances can occur where two basic network topologies, when connected together, can still retain the basic network character, and therefore not be a hybrid network. For example, a tree network connected to a tree network is

Figure 2.5: Ring Topology

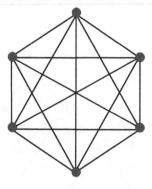

Figure 2.6: Fully Connected Topology

still a tree network. Therefore, a hybrid network accrues only when two basic networks are connected and the resulting network topology fails to meet one of the basic topology definitions. For example, two star networks connected together exhibit hybrid network topologies. A hybrid topology always accrues when two different basic network topologies are connected. See Figure 2.7.

Mesh topology: A network topology in which there are at least two nodes with two or more paths between them. See Figure 2.8.

Functions of Networks

With the evolution of networks, the functions of networks also extended consequently. The main function of networks can be identified as follows:

- **Information transmission**. The essential function of a network is to transfer information between a source and a destination. The communication may involve that transfer of a single block of information

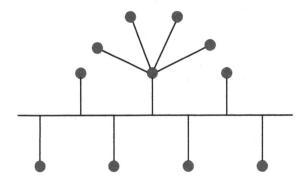

Figure 2.7: Hybrid Topology

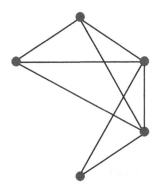

Figure 2.8: Mesh Topology

or the transfer of a stream of information between nodes in the network. The network must be able to provide connectivity in the sense of providing a means for information to flow among users. This basic capability is provided by transmission systems that transmit information by using various media such as wires, cables, radio, and optical fiber. Networks are typically designed to carry specific types of information representation, for example, analog voice signals, bits, or characters. Other function with transmission involves information representation; communication switching, information routing and forwarding; network addressing; communication traffic control; communication congestion control; and network management.

- **Information storage**. For the terminal users, networks are considered as a efficient media to exchange and share information. Thus more data, such as voice data, text, figures, multimedia data and potential new kind of data are hosted in networks. The research [How03] reveals that the size of the web information in the Internet got to 532,897 terabytes (TB) at the end of 2002. Thus the network becomes a massive data house and provides users with a variety of services.

- **Information process.** In the current era, networks can work as a basic platform for a business entity and for personal use. Networks cannot only play an important role for information storage, but also act as an crucial role in the information process with various computing techniques to meet the continuing requirements for network users. Heterogeneity networks provide the possibility that users can execute complex applications and obtain service from ubiquitous networks. Meanwhile, with the powerful computing capability, networks can work as a virtual society of real-life world, such as virtual university (education), virtual game environment, virtual social networks, etc.

- **Network Management.** The network operation must also ensure that network resources are used effectively under normal as well as under problem conditions. Traffic controls are necessary to ensure the smooth flow of information through the network. Network management functions includes monitoring the performance of the network, detecting and recovering from faults, configuring the network resources, maintaining accounting information for cost and billing purposes, and providing security by controlling access to the information flows in the network.

2.1.3 Types of Networks

Most common types of computer networks are identified, in order of scale, as following:

- **Personal Area Network (PAN)**

  A personal area network (PAN) is a computer network used for communication among computer devices close to one person. Some examples of devices that are used in a PAN are printers, fax machines, telephones, PDAs, and scanners. The reach of a PAN is typically about 20-30 feet (approximately 6-9 meters), but this is expected to increase with technology improvements.

- **Local Area Network (LAN)**

  This is a network covering a small geographic area, like a home, office, or building. Current LANs are most likely to be based on Ethernet technology. For example, a library may have a wired or wireless LAN for users to interconnect local devices (e.g., printers and servers) and to connect to the Internet. On a wired LAN, PCs in the library are typically connected by category 5 (Cat5) cable, running the IEEE 802.3 protocol through a system of interconnected devices and eventually connect to the Internet. The cables to the servers are typically on Cat 5e enhanced cable, which will support IEEE 802.3 at 1 Gbit/s. A wireless LAN may exist using a different IEEE protocol, 802.11b, 802.11g, or possibly 802.11n. The staff computers (bright green in the figure) can get to the color printer, checkout records and the academic network, and the Internet. All user computers can get to the Internet and the card catalog. Each workgroup can get to its local printer. Note that the printers are not accessible from outside their workgroup.

A typical library network is a branching tree topology with controlled access to resources. All interconnected devices must understand the network layer (layer 3), because they are handling multiple subnets. Those inside the library, which have only 10/100 Mbit/s Ethernet connections to the user device and a Gigabit Ethernet connection to the central router, could be called "layer 3 switches" because they only have Ethernet interfaces and must understand IP. It would be more correct to call them access routers, where the router at the top is a distribution router that connects to the Internet and academic networks' customer access routers.

The defining characteristics of LANs, in contrast to WANs (wide area networks), include their higher data transfer rates, smaller geographic range, and lack of a need for leased telecommunication lines. Current Ethernet or other IEEE 802.3 LAN technologies operate at speeds up to 10 Gbit/s. IEEE has projects investigating the standardization of 100 Gbit/s, and possibly 40 Gbit/s.

- **Campus Area Network (CAN)**

  This is a network that connects two or more LANs but that is limited to a specific and contiguous geographical area such as a college campus, industrial complex, office building, or a military base. A CAN may be considered a type of MAN (metropolitan area network), but is generally limited to a smaller area than a typical MAN. This term is most often used to discuss the implementation of networks for a contiguous area. This should not be confused with a Controller Area Network. A LAN connects network devices over a relatively short distance. A networked office building, school, or home usually contains a single LAN, though sometimes one building will contain a few small LANs (perhaps one per room), and occasionally a LAN will span a group of nearby buildings. In TCP/IP networking, a LAN is often but not always implemented as a single IP subnet.

- **Metropolitan Area Network (MAN)**

  A Metropolitan Area Network is a network that connects two or more Local Area Networks or Campus Area Networks together but does not extend beyond the boundaries of the immediate town/city. Routers, switches, and hubs are connected to create a Metropolitan Area Network.

- **Wide Area Network (WAN)**

  A WAN is a data communications network that covers a relatively broad geographic area (i.e., one city to another and one country to another country) and that often uses transmission facilities provided by common carriers, such as telephone companies. WAN technologies generally function at the lower three layers of the OSI reference model: the physical layer, the data link layer, and the network layer.

- **Global Area Network (GAN)**

  Global Area Networks (GAN) specifications are in development by several groups, and there is no common definition. In general, however, a GAN is a model for supporting mobile communications across an arbitrary number of wireless LANs, satellite coverage areas, etc. The key challenge in mobile communications is "handing off" the user communications from one local coverage area to the next. In IEEE Project 802, this involves a succession of terrestrial Wireless Local Area Networks (WLAN).

- **Internetwork**

  Two or more networks or network segments connected using devices that operate at layer 3 (the "network" layer) of the OSI Basic Reference Model, such as a router. Any interconnection among or between public, private, commercial, industrial, or governmental networks may also be defined as an internetwork.

  In modern practice, the interconnected networks use the Internet Protocol. There are at least three variants of internetwork, depending on who administers and who participates in them:

  - **Intranet**

    An intranet is a set of networks, using the Internet Protocol and IP-based tools such as web browsers and file transfer applications, that is under the control of a single administrative entity. That administrative entity closes the intranet to all but specific, authorized users. Most commonly, an intranet is the internal network of an organization. A large intranet will typically have at least one web server to provide users with organizational information.

  - **Extranet**

    An extranet is a network or internetwork that is limited in scope to a single organization or entity, but which also has limited connections to the networks of one or more other, usually but not necessarily, trusted organizations or entities (e.g., a company's customers may be given access to some part of its intranet creating in this way an extranet, while at the same time the customers may not be considered "trusted" from a security standpoint). Technically, an extranet may also be categorized as a CAN, MAN, WAN, or other type of network, although, by definition, an extranet cannot consist of a single LAN; it must have at least one connection with an external network.

– **Internet**

The Internet is a specific internetwork. It consists of a worldwide interconnection of governmental, academic, public, and private networks based upon the networking technologies of the Internet Protocol Suite. It is the successor of the Advanced Research Projects Agency Network (ARPANET) developed by DARPA of the U.S. Department of Defense. The Internet is also the communications backbone underlying the World Wide Web (WWW). The "Internet" is most commonly spelled with a capital "I" as a proper noun, for historical reasons and to distinguish it from other generic internetworks.

Participants in the Internet use a diverse array of methods of several hundred documented, and often standardized, protocols compatible with the Internet Protocol Suite and an addressing system (IP Addresses) administered by the Internet Assigned Numbers Authority and address registries. Service providers and large enterprises exchange information about the reachability of their address spaces through the Border Gateway Protocol (BGP), forming a redundant worldwide mesh of transmission paths.

Intranets and extranets may or may not have connections to the Internet. If connected to the Internet, the intranet or extranet is normally protected from being accessed from the Internet without proper authorization. The Internet is not considered to be a part of the intranet or extranet, although it may serve as a portal for access to portions of an extranet.

2.2   History of Networks

2.2.1   History of Telecommunications Networks

In ancient times, the most common way of producing a signal would be through light (fires) and sound (drums and horns).

1791:   The Chappe brothers in France set up a semaphore system to send messages to each other.

1793:   The Chappe brothers established the first commercial semaphore system between two locations near Paris.

1800: The starting point of all modern telecommunications was the invention of the electric cell by Alessandro Volta.

1809: Thomas S. Sommering proposed a telegraphic system.

1843: Samuel Morse proposed a way to assign each letter and number to a ternary code. In this year, Alexander Bain invented FAX.

1850: The transducer was invented to transform an acoustic signal into an electric one and vice versa (microphone and receiver) with acceptable information loss.

1867: The first Atlantic cable, promoted by Cyrus Field, was laid on July 27th.

1870: Thomas Edison invented multiplex telegraphy.

1876: Alexander Graham Bell invents the telephone.

1877: Western Union put first telephone line in operation between Somerville, MA, and Boston.

1878: Bell formed the Bell Telephone Company and established the first switching office in New Haven, CT, U.S.

1882: Bell had controlling interest in Western Union and Western Electric.

1887: Charles Vernon Boys described concept of guiding light through glass fibers.

1895: Guglielmo Marconi developed the first wireless telegraph system.

1899: Almon Strowger invented an electro-mechanic device known as "selector", which was directed by the electrical signals coming from the calling telephone device, achieved through selection based on geographical prefixes.

1910: Peter DeBye in Holland developed the theory for optical waveguides.

1913: AT&T agreed to divest its holdings of Western Union, stop acquisition of other telcos, and permitted other telcos to interconnect.

1914: Underground cables linked Boston, New York City, and Washington.

1915: Vacuum tube amplifiers used the first time in coast-to-coast telco circuits.

1920: Valve amplifiers made their first appearance.

1923: The television was invented.

1925: Bell Telephone Laboratories was founded.

1926: First public crossbar switch exchange opened in Sweden.

1927: First commercial radio telephone service operated between Britain and the U.S.

1934: Federal Communications Commission (FCC) founded.

1935: First telephone call around the world. About 6700 telcos in operation.

1938: Bell introduced crossbar central office switches.

1946: The invention of ENIAC (Electronic Numerical Integrator and Computer) started the era of informatics.

1947: The invention of transistors gave birth to the field of electronics.

1955: Modem first described by Ken Krechmer, A. W. Morten, and H. E. Vaughn.

1958: (1) The first integrated circuit was built. (2) AT&T introduces datasets (modems) for direct connection.

1959: AT&T introduced the TH-1 1860-channel microwave system.

1960: AT&T installed first electronic switching system.

1961: Bell Telephone Labs released design information for touch-tone dial to Western Electric.

1962: (1) AT&T introduced T-1 multiplex service in Skokie, Illinois. (2) Paul Baron introduced idea of distributed packet-switching networks. (3) The first communication satellite, Telstar, launched into orbit.

1967: Larry Roberts at the Advanced Research Projects Agency published a paper proposing ARPANET.

1969: The first microprocessor was invented.

1974: First domestic satellites in operation.

1975: Fiber optics were installed in U.S. and Europe.

1976: Digital radio and time division switching were introduced.

1977: The Advanced Mobile Phone System (AMPS), invented by Bell Labs, first installed in the U.S. with geographic regions divided into "cells" (i.e., cellular telephone).

1981: Bell Labs designed network-embedded database of Personal Identification Numbers (PINs) for calling card customers to be accessed by public telephones.

1983: TCP/IP was selected as the official protocol for the ARPANET.

1986: (1) The National Science Foundation of U.S. introduced its 56kbps backbone network. (2) The Joint Photographic Expert Group (JPEG) was founded by ITU, ISO, and IEC.

1987: Bellcore introduces Asymmetric Digital Subscriber Line (ADSL) concept, which has potential of multimedia transmission over nation's copper loops.

1988: (1) Telecommunications management network was produced by the International Telecommunication Union (ITU-T) as a strategic goal to create or identify standard interfaces that would allow a network to be managed consistently across all network element suppliers. (2) E.212 described a system to identify mobile devices as they move from network to network.

1989: SDH key standard was introduced for digital information over optical fiber.

1990: Recommendation H.261 (p × 64) video coding was introduced.

1992: (1) Bell Labs demonstrated 5-Gbps transmission of optical solitons over 15,000 km, and 10-Gbps over 11,000 km. (2) One-millionth host connected to the Internet.

1993: (1) The NSF network backbone jumped from T-1 to T-3. (2) The first DSL standard was consented.

1994: FCC licensed the Personal Communication Services (PCS) spectrum (1.7 to 2.3 GHz) for $7.7B.

1995: Nationwide Caller ID implemented.

1996: (1) H.323 Key facilitator was introduced for video-conference and VoIP. (2) UIFN (universal international freephone numbers) was adopted.

1997: (1) Lucent announced development of wireless loops with 128K ISDN capability. (2)New international telephone numbering plan - E.164 was produced.

1998: (1) Sprint Corp. announced an offer for an advanced packet-switching network to simultaneously send voice, data, and video down a single phone line. (2) Ericsson, IBM, Intel, Nokia, and Toshiba announced the development of Bluetooth for wireless data exchange between handheld computers or cellular phones and stationary computers.

1999: J.117 Key CableTV standard was introduced.

2000: Bearer independent call control was introduced.

2002: ITU-T Recommendation H.264 was introduced for advanced video coding for generic audiovisual services.

2003: ITU-T Recommendation H.350 was introduced for directory services architecture for multimedia Conferencing.

2004: NGN Focus Group was formed to smooth transition from PSTN to packet-based networks.

2005: VDSL2 further extended the use of legacy copper cabling and will be next important broadband technology.

2006: (1) The three leading wireless carriers rolled out 3G networks supporting data rates from 400K to 700K. (2) RFID, IPTV Focus Group was formed.

2007: Apple iPhone, displayed with multi-touch function, iPod video player, mobile phone, camera and Internet browser in one device.

2008: Apple iPhone 3G delivers UMTS, HSDPA, GSM, Wi-Fi, EDGE, GPS, and Bluetooth 2.0 + EDR in one compact device - using only two antennas.

2.2.2 History of Computer Networks (Internet)

The history of computer networks is actually parallel with the evolution of Internet [Zak][1] [Kle08] [ISC] [IWS]:

1961: First paper on packet-switching theory by Leonard Kleinrock.

1964: Paul Baran, RAND: "On Distributed Communications Networks" proposed Packet-switching networks.

1966: Lawrence G. Roberts, MIT: "Towards a Cooperative Network of Time-Shared Computers" presented the first ARPANET plan.

---

[1]Hobbes' Internet Timeline (c) Robert H Zakon, www.Zakon.org

1969:   (1) Birth of computer network: ARPANET commissioned by DoD for research into networking, and first 4 nodes were connected as a network. (2) Steve Crocker established the Request For Comments (RFC) series and authored the first RFC entitled "Host Protocol."

1970:   (1) The ARPANET spanned the United States with a connection from UCLA to BBN. (2) The Network Working Group (NWG) released the first host-to-host protocol called the Network Control Program (NCP). It was the first transport layer protocol of the ARPANET, later to be succeeded by TCP. (3) Norm Abramson developed Alohanet in Hawaii, a 9600-bps packet radio net based on the ALOHA multi-access technique of random access.

1972:   (1) Ray Tomlinson of BBN introduced network email and the @ sign. (2) First computer-to-computer chat takes place at UCLA, and was repeated during ICCC, as psychotic PARRY (at Stanford) discussed its problems with the Doctor (at BBN).

1973:   (1) First international connections to the ARPANET: University College of London (England) via NORSAR(Norway). (2) The first conception of Ethernet was made by Robert Metcalfe. (3) The Packet Satellite Net (SATNET) was attached to the ARPANET, based on a shared 64kb/s Intelsat IV channel. This was the first international connection and initially connected the United States and the United Kingdom. There were now three networks interconnected. (4) Bob Metcalfe invented Ethernet when he proposed the technology in a memo circulated at the Xerox Research Center in Palo Alto.

1974:   (1) TCP, or Transmission Control Program was first introduced. (2) First Use of term Internet by Vint Cerf and Bob Kahn in paper on Transmission Control Protocol.

1975:   (1) Telnet was developed. (2) Management of the ARPANET was transferred to the Defense Communications Agency (DCA).

1976:   (1) The Department of Defense began to experiment with the TCP/IP protocol and soon decided to require it for use on ARPANET. (2) X.25 protocols developed for public packet networking.

1977:   TCP was used to connect three networks (ARPANET, PRNET, and SATNET) in an intercontinental demonstration.

1978:   TCP split into TCP (Transmission Control Protocol) and IP (Internet Protocol).

1979:   (1) USENET established using UUCP between Duke and UNC by Tom Truscott, Jim Ellis, and Steve Bellovin. (2) CSNET was conceived as a result of a meeting convened by Larry Landweber. The National Science Foundation (NSF) funded it in early 1981. This enabled the connection of many more computer science researchers to the growing Internet.

1980: ARPANET ground to a complete halt on 27 October because of an accidentally propagated status-message virus. (2) Ethernet went commercial through 3-Com and other vendors. (3) IBM introduced their first personal computer (PC).

1982: Norway made network to become an Internet connection via TCP/IP over SATNET; UCL did the same.

1983: (1) The TCP/IP protocol was used in the ARPANET.

1984: The ARPANET was divided into two networks: MILNET and ARPANET. MILNET was to serve the needs of the military and ARPANET to support the advanced research component, Department of Defense continued to support both networks. (2) The Domain Name System (DNS) was designed by Paul Mockapetris.

1985: (1) The United States National Science Foundation initiated the development of NSFNET, which marked the birth of the Internet. (2) Whole Earth 'Lectronic Link (WELL) started.

1986: (1)The Internet Engineering Task Force or IETF was created to serve as a forum for technical coordination by contractors for DARPA working on ARPANET, U.S. Defense Data Network (DDN), and the Internet core gateway system. (2) Internet Engineering Task Force (IETF) and Internet Research Task Force (IRTF) came into existence under the IAB. First IETF meeting held in January at Linkabit in San Diego. (3) Network News Transfer Protocol (NNTP) designed to enhance Usenet news performance over TCP/IP.

1987: NSF signed a cooperative agreement to manage the NSFNET backbone with Merit Network, Inc. (IBM and MCI involvement was through an agreement with Merit). Merit, IBM, and MCI later founded ANS.

1988: (1) Internet worm burrowed through the Net, affecting about 6,000 of the 60,000 hosts on the Internet. (2) NSFNET backbone upgraded to T1 (1.544Mbps). (3) DoD chose to adopt OSI and used TCP/IP as an interim. U.S. Government OSI Profile (GOSIP) defined the set of protocols to be supported by Government purchased products. (3) Robert Morris unleashed the first Internet worm. This was the commencement of the dark side of the Internet.

1989: (1) RIPE (Reseaux IP Europ'eens) formed (by European service providers) to ensure the necessary administrative and technical coordination to allow the operation of the pan-European IP Network. (2) Corporation for Research and Education Networking (CREN) was formed by merging CSNET into BITNET. (3) ARPANET backbone replaced by NSFNET.

1990: (1) The World came on-line (world.std.com), becoming the first commercial provider of Internet dial-up access. (2) ISO Development Environment (ISODE) developed to provide an approach for OSI migration for the DoD. ISODE software allowed OSI application to operate over TCP/IP.

1991: (1) CSNET (which consisted of 56Kbps lines) was discontinued having fulfilled its important early role in the provision of academic networking service. (2) The NSF established a new network, named NREN, the National Research and Education Network. The purpose of this network was to conduct high-speed networking research. It was not to be used as a commercial network, nor was it to be used to send a lot of the data that the Internet now transfers. (3) World Wide Web (WWW) released by CERN; Tim Berners-Lee developer. (4) PGP (Pretty Good Privacy) released by Philip Zimmerman. (5) NSF acceptable use policies were changed to allow commercial traffic on the Internet.

1992: (1) Internet Society was chartered in January. (2) IAB reconstituted as the Internet Architecture Board and became part of the Internet Society.

1993: InterNIC created by NSF to provide specific Internet services: directory and database services (by AT&T), registration services (by Network Solutions Inc.), and information services (by General Atomics/CERFnet). The Mosaic browser was released by Marc Andreessen and Eric Bina of the National Center for Supercomputer Applications (NCSA) at the University of Illinois, Urbana-Champaign.

1994: (1) Pizza Hut offered pizza ordering on its Web page. (2) First Virtual, the first cyberbank, opened up for business. (3) ATM (Asynchronous Transmission Mode, 145Mbps) backbone was installed on NSFNET. (4) Netscape browser is released.

1995: (1) NSFNET reverted back to a research network. (2) Hong Kong police disconnected all but 1 of the colony's Internet providers in search of a hacker. 10,000 people were left without Net access. (3) RealAudio, an audio streaming technology, let the Net hear in near real-time. (4) Traditional online dial-up systems (Compuserve, America Online, Prodigy) began to provide Internet access (5) Netscape went public and the dot com boom started with the belief that a "new economy" was beginning. (6) Bill Gates issued "The Internet Tidal Wave" memo within Microsoft.

1996: (1) Internet phones caught the attention of U.S. telecommunication companies who asked the U.S. Congress to ban the technology. (2) New York's Public Access Networks Corp (PANIX) was shut down after repeated SYN attacks by a cracker using methods outlined in a hacker magazine (2600).

1997: (1) The American Registry for Internet Numbers (ARIN) was established to handle administration and registration of IP numbers to the geographical areas currently handled by Network Solutions (InterNIC), starting March 1998. (2) Internet2 consortium was established. (3) IEEE released 802.11 (WiFi) standard. (4) Barry Leiner et al., published a paper on "A Brief History of the Internet."

1998: (1) U.S. Department of Commerce (DoC) released the Green Paper outlining its plan to privatize DNS on 30 January. (2) Blogs began to appear. (3) Voice over IP (VoIP) equipment began rolling out.

1999: (1) First Internet Bank of Indiana, the first full-service bank available only on the Net, opened for business on 22 February. (2) IBM became the first Corporate partner to be approved for Internet2 access. (3) ICANN announced the five testbed registrars for the competitive Shared Registry System on 21 April. (4) First large-scale Cyberwar took place simultaneously with the war in Serbia/Kosovo. (5) Abilene, the Internet2 network, reached across the Atlantic and connected to NORDUnet and SURFnet. (6) Napster rolled out.

2000: (1) The U.S. timekeeper (USNO) and a few other time services around the world reported the new year as 19100 on 1 Jan. (2) Dot-com bubble began to burst. (3) A massive denial of service attack was launched against major web sites, including Yahoo, Amazon, and eBay in early February.

2001: (1) European Council finalized an international cybercrime treaty on 22 June and adopted it on 9 November. This was the first treaty addressing criminal offenses committed over the Internet. (2) Code Red worm and Sircam virus infiltrated thousands of web servers and e-mail accounts, respectively, causing a spike in Internet bandwidth usage and security breaches. (3) GÉANT, the pan-European Gigabit Research and Education Network, became operational, replacing the TEN-155 network which was closed down. (4) English was no longer the language of the majority of Internet users. It fell to a 45 percent share.

2002: (1) Global Terabit Research Network (GTRN) was formed composed of two OC-48 2.4GB circuits connecting Internet2 Abiline, CANARIE CA*net3, and GÉANT. (2) Abilene (Internet2) backbone deployed native IPv6.

2003: (1) Last Abilene segment upgraded to 10Gbps. (2) The World Wide Web Consortium announced its formal policy for ensuring that key web technologies, even if patented, were made available on a royalty-free basis. (3) Spam is e-mail that was not wanted, is one of the shortest definitions.

2004: (1) 250,000 computers infested by Mydoom. (2) Like MS explorer Google dominated the web in searching. The verb "to Google" became an accepted term. (3) Spam took up more than 33% of e-mail traffic and this number is growing fast. (4) U.S. mobile phone revenue of $50 billion equaled that of U.S. fixed-line phone revenue. (5) Camera-enabled phone sales exceeded combined sales of digital plus film cameras.

2005: (1) According to Netcraft there were approximately 75 million websites online. Over 17 million sites were added to the web this year. (2) Google was the darling of the Internet. (3) Peer-to-peer networks grew. Supreme Court decision supported Recording Industry Association of America (RIAA) position. (4) Google Maps and Google Earth appeared. (5) Web 2.0 technologies heated up. (6) MySpace had more page views than Google.

2006: (1) YouTube was purchased by Google. (2) Web 2.0 got popular and lead to high loads in the network.

2007: (1) Mobile TV ads, applications, and content emerged. (2) Apple introduced the iPhone. (3) Microsoft bought into Facebook at a $15 billion valuation. (4) Google laid out Android, its open cell phone platform. (5) WLAN standard 802.11n (540Mbps, 2.4GHz) was launched.

2008: (2) Intel launched the Atom platform devices MID (Mobile Internet Devices). (3) End of 2008, the number of Internet users gets to over 1,574,313,184 persons worldwide. The Internet Penetration Rate is 23.5%, considering a global population of 6,708,755,756 persons according to the U.S. Census Bureau data. (4) Google G1: 1 Smart phone with Android as the operating system

2009: The total number of Internet users for June 30, 2009 is estimated to be 1,668,870,408. This represents a 24.7% penetration rate worldwide, considering a global population of 6,767,805,208 persons.

Table 2.1 [Zak], Figure 2.9 and Figure 2.10 illustrate the growth of Internet hosts, web hosts, and Internet users over last 40 years. Figure 2.11 and 2.12 denote Internet users and Internet penetration rate by geographic regions in the world.

From the history of the Internet, the evolution of the Internet over last 4 decades demonstrate:

- Internet has grown from an experimental research network to an infrastructure supporting the economy as well as the provision of societal services.

- Internet is becoming pervasive and ubiquitous, with already 25% of the world population having access to it whilst mobile broadband is expected to rapidly provide access to another 2 Billions users globally.

Table 2.1: Statistics of Internet Hosts and Web Hosts

Year	Internet Hosts	Year	Internet Hosts	Web Hosts	Year	Internet Hosts	Web Hosts
1969	4	1983	562		1997	26053000	1681868
1970	13	1984	1024		1998	36739000	3518158
1971	23	1985	1961		1999	56218000	9560688
1972	31	1986	5089		2000	93047785	25675581
1973	35	1987	28174		2001	125888197	36276252
1974	62	1988	33000		2002	162128493	35543105
1975	n/a	1989	56000		2003	171638297	45980112
1976	n/a	1990	313000		2004	285139107	56923737
1977	111	1991	617000		2005	353284187	74353258
1978	n/a	1992	1136000	26	2006	439286364	105244649
1979	188	1993	2056000	623	2007	489774269	152604741
1980	204	1994	3864000	13900	2008	570937778	n/a
1981	213	1995	8200000	23500			
1982	235	1996	16729000	603367			

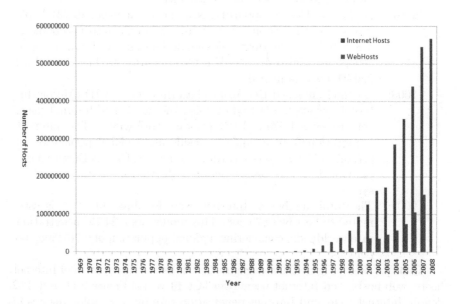

Figure 2.9: The Growth of Internet Hosts and Web Hosts

- Internet has enabled user and consumer empowerment, through the emergence of e-Commerce and social networks.

- Internet has helped the modernization of public administration through emergence of e-Government, e-Health, e-Education.

- Internet use is also expected to contribute significantly to solve emerging

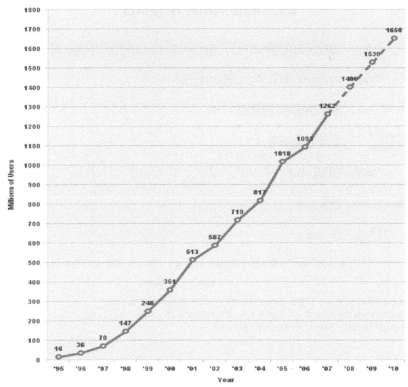

Source: www.internetworldstats.com - January, 2008
Copyright © 2008, Miniwatts Marketing Group

Figure 2.10:   The Growth of Internet Users

challenges such as climate change and energy efficiency.

- Internet has catalyzed innovation and favored the emergence of new disruptive business models: in 2008, about 300 million persons use free VoIP Skype Software; novel video consumption models such as YouTube have emerged, with global users uploading some 10 hours of video per minute from that site.

- Internet has supported the creativity of entrepreneurs through openness, which has made it possible for thousands of innovators worldwide, to develop applications acquiring almost instantly value through the networking multiplier effect.

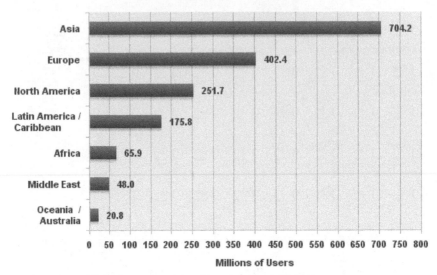

Figure 2.11: Internet Users in the World by Geographic Regions

2.3 Network Architectures

There are two important network architectures, the OSI reference model and the TCP/IP reference model.

2.3.1 The OSI Reference Model

The Open Systems Interconnection Reference Model (OSI Reference Model or OSI Model) is an abstract description for layered communications and computer network protocol design. It was developed as part of the Open Systems Interconnection (OSI) initiative. In its most basic form, it divides network architecture into seven layers which, from top to bottom, are the Application, Presentation, Session, Transport, Network, Data-Link, and Physical Layers. It is therefore often referred to as the OSI Seven Layer Model. See Figure 2.13.

A layer is a collection of conceptually similar functions that provide services to the layer above it and receives service from the layer below it. For example, a layer that provides error-free communications across a network

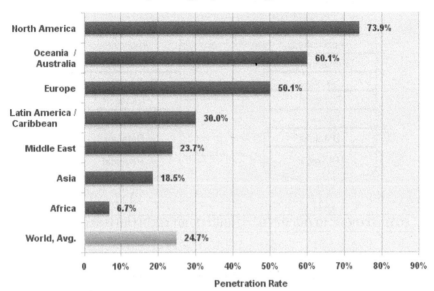

Figure 2.12: World Internet Penetration Rates by Geographic Regions

provides the path needed by applications above it, while it calls the next lower layer to send and receive packets that make up the contents of the path.

- Layer 7: Application Layer

  The application layer is the OSI layer closest to the end user, which means that both the OSI application layer and the user interact directly with the software application. This layer interacts with software applications that implement a communicating component. Such application programs fall outside the scope of the OSI model. Application layer functions typically include identifying communication partners, determining resource availability, and synchronizing communication. When identifying communication partners, the application layer determines the identity and availability of communication partners for an application with data to transmit. When determining resource availability, the application layer must decide whether sufficient network resources for the requested communication exist. In synchronizing communica-

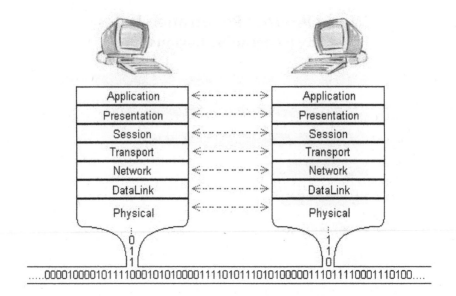

... 0000100001011110001010100001111010111010100001110111100011101100 ....

Figure 2.13:   OSI Reference Model

tion, all communication between applications requires cooperation that
is managed by the application layer. Some examples of application layer
implementations include Telnet, File Transfer Protocol (FTP), and Simple Mail Transfer Protocol (SMTP).

- Layer 6: Presentation Layer

  The Presentation Layer establishes a context between Application Layer
  entities, in which the higher-layer entities can use different syntax and
  semantics, as long as the Presentation Service understands both and the
  mapping between them. The presentation service data units are then
  encapsulated into Session Protocol Data Units, and moved down the
  stack. This layer provides independence from differences in data representation (e.g., encryption) by translating from application to network
  format, and vice versa. The presentation layer works to transform data
  into the form that the application layer can accept. This layer formats
  and encrypts data to be sent across a network, providing freedom from
  compatibility problems. It is sometimes called the syntax layer.

- Layer 5: Session Layer

  The Session Layer controls the dialogues/connections (sessions) between
  computers. It establishes, manages, and terminates the connections
  between the local and remote application. It provides for full-duplex,
  half-duplex, or simplex operation, and establishes checkpointing, ad-

journment, termination, and restart procedures. The OSI model made this layer responsible for "graceful close" of sessions, which is a property of TCP, and also for session checkpointing and recovery, which is not usually used in the Internet Protocol Suite. The Session Layer is commonly implemented explicitly in application environments that use remote procedure calls (RPCs).

- Layer 4: Transport Layer

The Transport Layer provides transparent transfer of data between end users, providing reliable data transfer services to the upper layers. The Transport Layer controls the reliability of a given link through flow control, segmentation/desegmentation, and error control. Some protocols are state and connection oriented. This means that the Transport Layer can keep track of the segments and retransmit those that fail. The best known examples of a Layer 4 protocol are the Transmission Control Protocol (TCP) and User Datagram Protocol (UDP).

- Layer 3: Network Layer

The Network Layer provides the functional and procedural means of transferring variable length data sequences from a source to a destination via one or more networks, while maintaining the quality of service requested by the Transport Layer. The Network Layer performs network routing functions, and might also perform fragmentation and reassembly, and report delivery errors. Routers operate at this layer: sending data throughout the extended network and making the Internet possible. This is a logical addressing scheme: values are chosen by the network engineer. The addressing scheme is hierarchical. The best-known example of a Layer 3 protocol is the Internet Protocol (IP).

- Layer 2: Data Link Layer

The Data Link Layer provides the functional and procedural means to transfer data between network entities and to detect and possibly correct errors that may occur in the Physical Layer. Originally, this layer was intended for point-to-point and point-to-multipoint media, characteristic of wide area media in the telephone system. Local area network architecture, which included broadcast-capable multiaccess media, was developed independently of the ISO work, in IEEE Project 802.

- Layer 1: Physical Layer

The Physical Layer defines the electrical and physical specifications for devices. In particular, it defines the relationship between a device and a physical medium. This includes the layout of pins, voltages, cable specifications, Hubs, repeaters, network adapters, Host Bus Adapters (HBAs used in Storage Area Networks), and more.

The major functions and services performed by the Physical Layer are establishment and termination of a connection to a communications medium. Participation in the process whereby the communication resources are effectively shared among multiple users. For example, contention resolution and flow control.

2.3.2 The TCP/IP Reference Model

The TCP/IP model and related protocols are maintained by the Internet Engineering Task Force (IETF).

The TCP/IP Suite defines a set of rules to enable computers to communicate over a network. TCP/IP provides end-to-end connectivity specifying how data should be formatted, addressed, shipped, routed, and delivered to the right destination. The specification defines protocols for different types of communication between computers and provides a framework for more detailed standards.

TCP/IP is generally described as having four abstraction layers: Application Layer, Internet Layer, Transport Layer, and Link Layer. See Figure 2.14. This layer view is often compared with the seven-layer OSI Reference Model formalized after the TCP/IP specifications.

- Application Layer

  The Application Layer refers to the higher-level protocols used by most applications for network communication. Examples of application layer protocols include the File Transfer Protocol (FTP) and the Simple Mail Transfer Protocol (SMTP). Data coded according to application layer protocols are then encapsulated into one or (occasionally) more transport layer protocols, such as the Transmission Control Protocol (TCP) or User Datagram Protocol (UDP), which in turn use lower-layer protocols to effect actual data transfer.

  Since the IP stack defines no layers between the application and transport layers, the application layer must include any protocols that act like the OSI's presentation and session-layer protocols. This is usually done through libraries.

  Application Layer protocols generally treat the transport layer (and lower) protocols as "black boxes" which provide a stable network connection across which to communicate, although the applications are usually aware of key qualities of the transport layer connection such as the end point IP addresses and port numbers.

- Transport Layer

  The Transport Layer's responsibilities include end-to-end message transfer capabilities independent of the underlying network, along with error

Network Connections

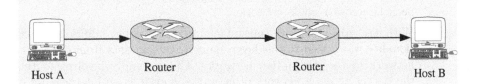

Stack Connections

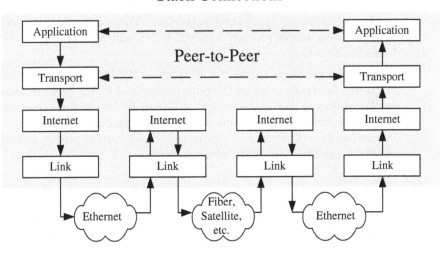

Figure 2.14: TCP/IP Reference Model

control, fragmentation, and flow control. End-to-end message transmission or connecting applications at the transport layer can be categorized as either: connection-oriented (e.g., TCP) or connectionless (e.g., UDP).

The Transport Layer can be thought of literally as a transport mechanism, e.g., a vehicle whose responsibility is to make sure that its contents (passengers/goods) reach its destination safely and soundly, unless a higher or lower layer is responsible for safe delivery.

The Transport Layer provides this service of connecting applications together through the use of ports. Since IP provides only a best-effort delivery, the Transport Layer is the first layer of the TCP/IP stack to offer reliability. Note that IP can run over a reliable data link protocol such as the High-Level Data Link Control (HDLC). Protocols above transport, such as RPC, also can provide reliability.

- Internet Layer

  The Internet layer (or Network Layer) solves the problem of getting packets across a single network.

  With the advent of the concept of internetworking, additional functionality was added to this layer, namely, getting data from the source network to the destination network. This generally involves routing the packet across a network of networks, known as an internetwork or internet (lower case).

  In the Internet Protocol Suite, IP performs the basic task of getting packets of data from source to destination. IP can carry data for a number of different upper-layer protocols. These protocols are each identified by a unique protocol number: ICMP and IGMP are protocols 1 and 2, respectively.

  Some of the protocols carried by IP, such as ICMP (used to transmit diagnostic information about IP transmission) and IGMP (used to manage IP Multicast data), are layered on top of IP but perform internetwork layer functions. This illustrates an incompatibility between the Internet and the IP stack and OSI model. Some routing protocols, such as OSPF, are also part of the network layer.

- Link Layer

  The Link Layer is the networking scope of the local network connection to which a host is attached. This regime is called the link in Internet literature. This is the lowest component layer of the Internet protocols, as TCP/IP is designed to be hardware independent. As a result TCP/IP has been implemented on top of virtually any hardware networking technology in existence.

  The Link Layer is used to move packets between the Internet Layer interfaces of two different hosts on the same link. The processes of transmitting packets on a given link and receiving packets from a link can be controlled both in the software device driver for the network card, as well as on firmware or specialist chipsets. These will perform data link functions such as adding a packet header to prepare it for transmission, then actually transmit the frame over a physical medium.

  The Link Layer can also be the layer where packets are intercepted to be sent over a virtual private network or other networking tunnel. When this is done, the Link Layer data are considered as application data and proceed back down the IP stack for actual transmission. On the receiving end, the data go up through the IP stack twice (once for routing and the second time for the tunneling function). In these cases a transport protocol or even an application scope protocol constitutes a virtual link placing the tunneling protocol in the Link Layer of the protocol stack.

Thus, the TCP/IP model does not dictate a strict hierarchical encapsulation sequence and the description is dependent upon actual use and implementation.

2.3.3   Comparison of OSI Model and TCP/IP Model

In the TCP/IP model of the Internet, protocols are deliberately not as rigidly designed into strict layers as the OSI model. However, TCP/IP does recognize four broad layers of functionality, which are derived from the operating scope of their contained protocols, namely, the scope of the software application, the end-to-end transport connection, the internetworking range, and lastly the scope of the direct links to other nodes on the local network.

Even though the concept is different than in OSI, these layers are nevertheless often compared with the OSI layering scheme in the following way:

The Internet Application Layer includes the OSI Application Layer, Presentation Layer, and most of the Session Layer. Its end-to-end Transport Layer includes the graceful close function of the OSI Session Layer as well as the OSI Transport Layer.

The internetworking layer (Internet Layer) is a subset of the OSI Network Layer, while the Link Layer includes the OSI Data Link and Physical Layers, as well as parts of OSI's Network Layer. These comparisons are based on the original seven-layer protocol model as defined in ISO, rather than refinements in such things as the internal organization of the Network Layer document.

The presumably strict consumer/producer layering of OSI as it is usually described does not present contradictions in TCP/IP, as it is permissible that protocol usage does not follow the hierarchy implied in a layered model. Such examples exist in some routing protocols (e.g., OSPF), or in the description of tunneling protocols, which provide a Link Layer for an application, although the tunnel host protocol may well be a Transport or even an Application Layer protocol in its own right.

The TCP/IP design generally favors decisions based on simplicity, efficiency, and ease of implementation.

2.3.4   Evolution of the Internet Protocol (IP)

The Internet Protocol (IP) is a protocol used for communicating data across a packet-switched internetwork using the Internet Protocol Suite, also referred to as TCP/IP.

IP is the primary protocol in the Internet Layer of the Internet Protocol Suite and has the task of delivering distinguished protocol datagrams (packets) from the source host to the destination host solely based on their addresses. For this purpose the Internet Protocol defines addressing methods and structures for datagram encapsulation. The first major version of addressing structure, now referred to as Internet Protocol Version 4 (IPv4), is still the dominant protocol of the Internet, although the successor, Internet Protocol Version 6 (IPv6), is being actively deployed worldwide.

Data from an upper layer protocol is encapsulated as packets/datagrams (the terms are basically synonymous in IP). Circuit setup is not needed before a host may send packets to another host that it has previously not communicated with (a characteristic of packet-switched networks), thus IP is a connectionless protocol. This is in contrast to Public Switched Telephone Networks that require the setup of a circuit before a phone call may go through (connection-oriented protocol).

Because of the abstraction provided by encapsulation, IP can be used over a heterogeneous network, i.e., a network connecting computers may consist of a combination of Ethernet, ATM, FDDI, Wi-Fi, token ring, or others. Each link layer implementation may have its own method of addressing (or possibly the complete lack of it), with a corresponding need to resolve IP addresses to data link addresses. This address resolution is handled by the Address Resolution Protocol (ARP) for IPv4 and Neighbor Discovery Protocol (NDP) for IPv6.

IPv4

Internet Protocol version 4 (IPv4) is the fourth revision in the development of the Internet Protocol (IP), and it is the first version of the protocol to be widely deployed. Together with IPv6, it is at the core of standards-based internetworking methods of the Internet and is still by far the most widely deployed Internet Layer protocol.

It is described in IETF publication RFC[2] 791 (September 1981), which rendered obsolete RFC 760 (January 1980). The United States Department of Defense also standardized it as MIL-STD-1777.

IPv4 is a data-oriented protocol to be used on a packet switched internetwork (e.g., Ethernet). It is a best-effort delivery protocol in that it does not guarantee delivery, nor does it assure proper sequencing, or avoid duplicate delivery. These aspects are addressed by an upper-layer protocol (e.g., TCP, and partly by UDP). IPv4 does, however, provide data integrity protection through the use of packet checksums.

IPv4 uses 32-bit (four-byte) addresses, which limits the address space to 4,294,967,296 (2^{32}) possible unique addresses. However, some are reserved for special purposes such as private networks ($\sim$18 million addresses) or multicast addresses ($\sim$16 million addresses). This reduces the number of addresses that can be allocated as public Internet addresses. As the number of addresses available are consumed, an IPv4 address shortage appears to be inevitable, however Network Address Translation (NAT) has significantly delayed this inevitability.

This limitation has helped stimulate the push towards IPv6, which is currently in the early stages of deployment and is currently the only contender

---

[2]All RFCs listed in this book can be accessed from http://www.ietf.org/rfc.html

to replace IPv4.

IPv4 addresses are usually written in dot-decimal notation, which consists of the four octets of the address expressed in decimal and separated by periods, such as: 10.91.0.221.

IPv6

Internet Protocol version 6 (IPv6) is the next-generation Internet Layer protocol for packet-switched internetworks and the Internet. IPv4 is currently the dominant Internet Protocol version, and was the first to receive widespread use. In December 1998, the Internet Engineering Task Force (IETF) designated IPv6 as the successor to version 4 by the publication of a Standards Track specification, RFC 2460.

IPv6 has a much larger address space than IPv4. This is based on the definition of a 128-bit address, whereas IPv4 used only 32 bits. The new address space thus supports 2^{128} (about 3.4×10^{38}) addresses. This expansion provides flexibility in allocating addresses and routing traffic and eliminates the need for network address translation (NAT). NAT gained widespread deployment as an effort to alleviate IPv4 address exhaustion.

IPv6 also implements new features that simplify aspects of address assignment (stateless address autoconfiguration) and network renumbering (prefix and router announcements) when changing Internet connectivity providers. The IPv6 subnet size has been standardized by fixing the size of the host identifier portion of an address to 64 bits to facilitate an automatic mechanism for forming the host identifier from Link Layer media addressing information (MAC address).

Network security is integrated into the design of the IPv6 architecture. Internet Protocol Security (IPsec) was originally developed for IPv6, but found widespread optional deployment first in IPv4 (into which it was back-engineered). The IPv6 specifications mandate IPsec implementation as a fundamental interoperability requirement.

IPv6 addresses are normally written as eight groups of four hexadecimal digits, where each group is separated by a colon (:), such as 2001:0db8:85a3:0000:0000:8a2e:0370:7334.

IPv4 vs. IPv6

To a great extent, IPv6 is a conservative extension of IPv4. Most transport- and application-layer protocols need little or no change to work over IPv6; exceptions are applications protocols that embed network-layer addresses (such as FTP or NTPv3). IPv6 specifies a new packet format, designed to minimize packet-header processing. Since the headers of IPv4 and IPv6 are significantly different, the two protocols are not interoperable.

- Larger address space

IPv6 provides a straightforward and long-term solution to the address space problem. IPv6 allows every citizen, every network operator (including those moving to all IP-"Next Generation Networks"), and every organization in the world to have as many IP addresses as they need to connect every conceivable device or good directly to the global Internet. IPv6 features a larger address space than that of IPv4: addresses in IPv6 are 128 bits long versus 32 bits in IPv4.

The very large IPv6 address space supports a total of 2^{128} (about 3.4×10^{38}) addressesor approximately 5×10^{28} (roughly 2^{95}) addresses for each of the roughly 6.5 billion (6.5×10^{9}) people alive today. In a different perspective, this is 2^{52} addresses for every observable star in the known universe.

While these numbers are impressive, it was not the intent of the designers of the IPv6 address space to ensure geographical saturation with usable addresses. Rather, the longer addresses allow a better, systematic, hierarchical allocation of addresses and efficient route aggregation. With IPv4, complex Classless Inter-Domain Routing (CIDR) techniques were developed to make the best use of the small address space. Renumbering an existing network for a new connectivity provider with different routing prefixes is a major effort with IPv4, as discussed in RFC 2071 and RFC 2072. With IPv6, however, changing the prefix in a few routers can renumber an entire network ad hoc, because the host identifiers (the least-significant 64 bits of an address) are decoupled from the subnet identifiers and the network provider's routing prefix.

The size of a subnet in IPv6 is 2^{64} addresses (64-bit subnet mask); the square of the size of the entire IPv4 Internet. Thus, actual address space utilization rates will likely be small in IPv6, but network management and routing will be more efficient.

- Stateless address autoconfiguration

  IPv6 hosts can configure themselves automatically when connected to a routed IPv6 network using ICMPv6 router discovery messages. When first connected to a network, a host sends a link-local multicast router solicitation request for its configuration parameters; if configured suitably, routers respond to such a request with a router advertisement packet that contains network-layer configuration parameters.

  If IPv6 Stateless Address Auto Configuration (SLAAC) is unsuitable for an application, a host can use stateful configuration (DHCPv6) or be configured manually. Stateless autoconfiguration is not used by routers.

- Multicast

  Multicast, the ability to send a single packet to multiple destinations, is part of the base specification in IPv6. This is unlike IPv4, where it is optional (although usually implemented).

IPv6 does not implement broadcast, the ability to send a packet to all hosts on the attached link. The same effect can be achieved by sending a packet to the link-local all hosts multicast group.

Most environments, however, do not currently have their network infrastructures configured to route multicast packets; multicasting on single subnet will work, but global multicasting might not.

- Mandatory network layer security

  Internet Protocol Security (IPsec), the protocol for IP encryption and authentication, forms an integral part of the base protocol suite in IPv6. IPSec support is mandatory in IPv6; this is unlike IPv4, where it is optional (but usually implemented). IPsec, however, is not widely used at present except for securing traffic between IPv6 Border Gateway Protocol routers.

- Simplified processing by routers

  A number of simplifications have been made to the packet header, and the process of packet forwarding has been simplified, in order to make packet processing by routers simpler and hence more efficient. Concretely,

  (1) The packet header in IPv6 is simpler than that used in IPv4, with many rarely used fields moved to separate options; in effect, although the addresses in IPv6 are four times larger, the (option-less) IPv6 header is only twice the size of the (option-less) IPv4 header.

  (2) IPv6 routers do not perform fragmentation. IPv6 hosts are required to either perform PMTU discovery, perform end-to-end fragmentation, or to send packets smaller than the IPv6 minimum maximum packet size.

  (3) The IPv6 header is not protected by a checksum, integrity protection is expected to be ensured by a transport-layer checksum. In effect, IPv6 routers do not need to recompute a checksum when header fields (such as the TTL or Hop Count) change. This improvement may have been made obsolete by the development of routers that perform checksum computation at line speed using dedicated hardware.

  (4) The Time-to-Live field of IPv4 has been renamed to Hop Limit, reflecting the fact that routers are no longer expected to compute the time a packet has spent in a queue.

- Mobility

  Unlike mobile IPv4, Mobile IPv6 (MIPv6) avoids triangular routing and is therefore as efficient as normal IPv6. However, since neither MIPv6 nor MIPv4 are widely deployed today, this advantage is mostly theoretical.

- Options Extensibility

  IPv4 has a fixed size (40 bytes) of option parameters. In IPv6, options are implemented as additional extension headers after the IPv6 header, which limits their size only by the size of an entire packet.

- Jumbograms

  IPv4 limits packets to 64 KB of payload. IPv6 has optional support for packets over this limit, referred to as jumbograms, which can be as large as 4 GB. The use of jumbograms may improve performance over high-MTU networks. The presence of jumbograms is indicated by the Jumbo Payload Option header.

To conclude, the key advantage of IPv6 over IPv4 is the huge, more easily managed address space. This solves the future problem of address availability now and for a long time to come. It provides a basis for innovation, developing and deploying services and applications that may be too complicated or too costly in an IPv4 environment. It also empowers users, allowing them to have their own network connected to the Internet.

2.4 Future of Networks

2.4.1 Laws Related to Network Evolution

Three laws that are generally accepted as governing the spread of technology and are related to the evolution of networks:

- Moore's Law: formulated by Gordon Moore of Intel in the early 1970s – the processing power of a microchip doubles every 18 months; corollary, computers become faster and the price of a given level of computing power halves every 18 months. It describes a long-term trend in the history of computing hardware. Figure 2.15 illustrates the Moore's Law.

- Gilder's Law: proposed by George Gilder in 1997, prolific author and prophet of the new technology age - the total bandwidth of communication systems triples every twelve months for the next 25 years. Bandwidth grows at least three times faster than computer power. While computer power doubles every eighteen months (Moore's law), communications power doubles every six months. The cost per communication bit will begin to sink farther than it has fallen previously. Eventually the cost of a telephone call, or of a bit transmitted, will be "free." See Figure 2.16. New developments seem to confirm that bandwidth availability will continue to expand at a rate that supports Gilder's Law.

- Metcalfe's Law: attributed to Robert Metcalfe. It states that the value of a telecommunications network is proportional to the square of the

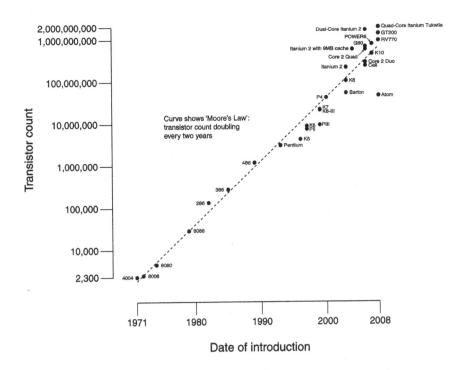

Figure 2.15: Moore's Law

number of connected users of the system (n^2). See Figure 2.17. Metcalfe's law characterizes many of the network effects of communication technologies and networks such as the Internet, social networking, and the World Wide Web. It is related to the fact that the number of unique connections in a network of a number of nodes (n) can be expressed mathematically as the triangular number $n(n-1)/2$, which is proportional to n^2 asymptotically.

Over 4 decades, the evolution of networks demonstrate the following facts:

- The scale of communication networks expanded from local telephone networks to GSM (Global System for Mobile communications), from the well-known infrastructure of cellular networks to non-infrastructure wireless ad-hoc networks and integrated with various wireless systems and Internet. The scale of computer networks evolves from ARPANET to LAN, PAN, WAN, GAN, pervasive Internet, and paralleled with the emerging of various ad hoc networks, even to nanonetworks.

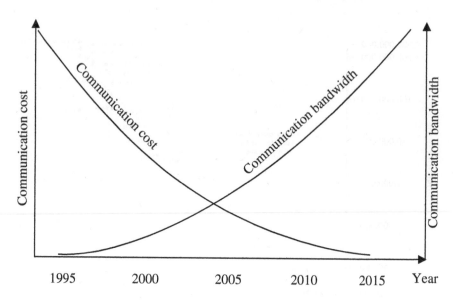

Figure 2.16: Gilder's Law

- The technology updating enables the network data traffic from a few k/s to over 10G/s, allows the network to transfer from plain text file to images, voice, video, and emerging rich media. Networks act as from simple message transformation to complex network services, such as e-commerce, e-government, e-learning, e-library, e-laboratory, e-health, e-society, online virtual reality, 3D Internet, etc.

It's difficult to give exact perdition about future networks. However, the ongoing evolution of networks indicates that networks will penetrate to various aspects of our real life and will bring tremendous changes in technology, research, industry, culture, military, and social life.

2.4.2   Trend of Networks

Future networks will probably have novel characteristics respect to today's networks. The technical and application trends of networks could be:

1. Nomadic computing (Mobility)

   Mobility in both the terminals and the services will have to be taken into consideration in future network designs. The number of mobile networked devices as well as nomadic users will increase dramatically. Subsequently more users and devices are connected and have direct dynamic communication link.

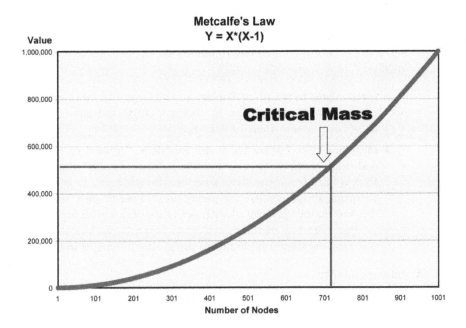

Figure 2.17: Metcalfe's Law

2. Wireless high-speed networks

The network applications are being complemented with really high-capacity and low-cost wireless access alternatives for finest possible access granularity and largest coverage for high speed access to networks. For example, next generation Mobile WiMAX network could transmit data at a speed of up to one gigabit per second while stationary, and 100 megabits per second in a moving vehicle. Current cellular technologies such as HSDPA have data speed of up to 5 megabits per second (in the downlink), and its expected to increase dramatically from current capacities in 3G and HSDPA towards HSDPA++, 3G LTE, 4G and beyond.

3. Scalability

The increasing scale of networks brings new challenges in a number of areas. Examples include modeling, validation, and verification of business processes composed on SOA; flexible evolution and execution of business processes; data, process, and service mediation; reliable management of composed services; and brokering, aggregation, and data management. Quality of software is an important factor in all of these and will become essential to the smooth working of the "service universe."

4. Security (Trust)

Security is becoming one most important footstone for modern network services. Creating trusted environments for the new service world will require:

- mechanisms to monitor, display, and analyze information flows between nodes participating in complex collaborations in order to detect and assess security risk; and

- mechanisms to ensure trust and confidence in services created by end-users themselves, i.e., built-in safeguards and guarantees so that others trust the new services. In addition, it is necessary to bring about changes in perception. Peer-to-peer services today are mainly associated with activities of doubtful legality, such as illegal trading of rights-protected content. Technical and legal mechanisms should be found to bring about changes in attitudes.

5. Interoperability

Network interoperability applies at many different levels:

- service interoperability to provide the ability to integrate largely stand-alone services with similar ones and with other services, for instance from the business domain;

- semantic interoperability, so as to provide the (automated) understanding of the information exchanged and ensure quality of service;

- interoperability of the service layer with network and application layers from different providers.

6. Context-awareness [Sch03]

The growing importance of context-awareness, targeting enriched experience, intuitive communications services fitting mobile lifestyle, and a mobilized workforce will in the future lead it to be more and more included in intelligent services that are smart but invisible to users. The social and economic benefits of making ICT-based services in areas as diverse as health, sustainable environment, safety, and transportation more intelligent and adaptive are recognized as a new driver for network services.

7. Autonomic computing [JL04]

Besides enhanced user experience for human-to-human or human-to-machine interactions, autonomous machine-to-machine communication has gained significant importance. More and more business transactions and processes will be automated and will take place based on autonomous decisions without any human intervention. These will be often based on or influenced by context information obtained from the physical world, without the requirement of human input to describe the situation. The emergence of the Web2.0 and associated technologies is just a starting point of this development, and already the impact of those on economic development is hugely beneficial.

Effectively, this enables an environment where real-world physical phenomena are electronically sampled and influenced by heterogeneous sensors and sensor/actuator islands and are at the fingertips of applications and humans alike, thus linking the physical world with the future networks. Consequently, our environment can be adjusted to our needs, or we can adjust our behavior following environmental changes. And our economic and social interactions are enhanced with efficient information or intelligent and autonomous machine-to-machine (M2M) interactions, enabling feedback and control loops, which are currently based on human input and which are cumbersome, slow, and fault ridden.

8. Integration

   The network will become increasingly integrated with phones, televisions sets, home appliances, portable digital assistants, and a range of other small hardware devices, providing an unprecedented, nearly uniform level of integrated data communications. Users will be able to access, status, and control this connected infrastructure from anywhere on the network.

9. Expanded Services

   Services (not only those for the end Users but also network services) are likely to be comprised of a variety of components, provided by a variety of Players (e.g., ASP, Prosumers) and running over a decentralized hosting (low-cost) infrastructure (including end-user devices, PC, servers, storage, computing and networking/forwarding resources, etc.). This vision is expected to pave the way for a deep integration of service and network frameworks for network convergence thus allowing broad federations of Players (e.g., Network and Service Providers and Application Service Providers) according to new business models. Openness, broad federations of Players, and do-it-yourself innovative services and knowledge management will allow people (already Prosumers as from Web2.0) to be the true center of Information Society.

Chapter Review

1. Will the the definition of network be improved with the trend of evolution of networks?

2. Why should we identify the topologies of networks?

3. Please list different standards for the classification of networks.

4. Please rethink of network history and network evolution.

5. What's the difference between OSI network architecture and Internet (TCP/IP) based network architecture? Why the latter surpasses the former in practical application?

6. What's the advantages of IPv6 over IPv4? What's the potential improvement for the future Internet protocol?

7. Predict the future network applications and challenges.

Chapter 3

Evolution in Network Management

3.1 Introduction of Network Management

3.1.1 Definition of Network Management

The definition of network management has different description, based on different points of view. Normally, network management is defined as the execution of the set of functions required for controlling, planning, allocating, deploying, coordinating, and monitoring the resources of a telecommunications network or a computer network, including performing functions such as initial network planning, frequency allocation, predetermined traffic routing to support load balancing, cryptographic key distribution authorization, configuration management, fault management, security management, performance management, and accounting management. Generally, network management does not include user terminal equipment.

Hegering [HAN99] defines network management as all measures ensuring the effective and efficient operations of a system within its resources in accordance with corporate goals. To achieve this, network management is tasked with controlling network resources, coordinating network services, monitoring network states, and reporting network status and anomalies. The objectives of network management are:

- **Managing system resources and services**: this includes control, monitor, update, and report of system states, device configurations, and network services.

- **Simplifying systems management complexity**: is the task of management systems that extrapolates systems management information into a humanly manageable form. Conversely, management systems should also have the ability to interpret high-level management objectives.

- **Providing reliable services**: means to provide networks with a high quality of service and to minimize system downtime. Distributed management systems should detect and fix network faults and errors. Network management must safeguard against all security threats.

- **Maintaining cost consciousness**: requires to keep track of system resources and network users. All network resource and service usage should be tracked and reported.

Another acceptable definition identifies network management as the activities, methods, procedures, and tools that pertain to the operation, administration, maintenance, and provisioning of networked systems [Cle06].

- Operation deals with keeping the network (and the services that the network provides) up and running smoothly. It includes monitoring the network to spot problems as soon as possible, ideally before users are affected.

- Administration deals with keeping track of resources in the network and how they are assigned. It includes all the "housekeeping" that is necessary to keep the network under control.

- Maintenance is concerned with performing repairs and upgrades – for example, when equipment must be replaced, when a router needs a patch for an operating system image, when a new switch is added to a network. Maintenance also involves corrective and preventive measures to make the managed network run "better," such as adjusting device configuration parameters.

- Provisioning is concerned with configuring resources in the network to support a given service. For example, this might include setting up the network so that a new customer can receive voice service.

In short, network management involves the planning, organizing, monitoring, accounting, and controlling of activities and resources and to keep the network service available and correct.

3.1.2 Basic Components of Network Management System

Network management has three main components: a managing center, a managed device, and a network management protocol.

The managing center consists of the network administrator and his or her facilities.

A managed device is the network equipment, including its software, that is controlled by the managing center. Any hub, bridge, router, server, printer, or modem can be a managed device.

The network management protocol is a policy between the managing center and the managed devices. The protocol in this context allows the managing center to obtain the status of managed devices.

Network management system contains two primary elements: a manager and agents.

The manager is the console through which the network administrator performs network management functions. A manager can be a network administrative device, as a management host.

Agents are the entities that interface to the actual device being managed. An agent can use the network management protocol to inform the managing center of an unexpected event. Bridges, hubs, routers or network servers are examples of managed devices that contain managed objects. These managed objects might be hardware, configuration parameters, performance statistics, and so on, that directly relate to the current operation of the device in question. These objects are arranged in what is known as a virtual information database, called a management information base, also called MIB. Network management protocols (such as SNMP, CMIP) allow managers and agents to communicate for the purpose of accessing these objects.

As specified in Internet RFCs and other documents, a typical distributed management system comprises:

- **Network elements**: Equipments which communicate with the network, according to standards defined by the ITU-T, with the purpose of being monitored or controlled, are named network elements. Sometimes they are also called managed devices [ITU96]. Network elements are hardware devices such as computers, routers, and terminal servers that are connected to networks. A network element is a network node that contains an SNMP agent, which resides on a managed network.

- **Manager**: A manager generates commands and receives notifications from agents. There are usually only a few managers in a system.

- **Agents**: Agents collect and store management information such as the number of error packets received by a network element. An agent has local knowledge of management information and transforms that information into the form compatible with SNMP. An agent responds to commands from the manager and sends notification to the manager. There are potentially many agents in a system.

- **Managed object**: A managed object is a vision of a feature of a network, from the point of view of the management system [ITU92]. All physical and logical resources, such as signaling terminals, routes, event logs, alarm reports and subscriber data, are regarded as managed objects. For example, in IP networks, a list of current active TCP circuits in a particular host computer is a managed object. Managed objects differ from variables, which are particular object instances. Managed objects can be scalar (defining a single object instance) or tabular (defining multiple and related instances). In literature, "managed object" is sometimes used interchangeably with "managed element."

- **Network Management Stations (NMSs)**: Sometimes NMSs are called consoles. These devices execute management applications that monitor and control network elements. Physically, NMSs are usually engineering workstation-caliber computers with fast CPUs, mega pixel color displays, substantial memory, and abundant disk space. At least one NMS must be present in each managed environment.

- **Management protocol**: A management protocol is used to convey management information between agents and network management stations (NMSs). Simple Network Management Protocol (SNMP) is the Internet community's de facto standard management protocol.

- **Structure of Management Information (SMI)**

  The structure of management information (SMI) language is used to define the rules for naming objects and to encode objects in a managed network center. In other words, SMI is a language by which a specific instance of the data in a managed network center is defined.

  SMI subdivides into three parts: module definitions, object definitions, and notification definitions.

  Module definitions are used when describing information modules. An ASN.1 macro, MODULE-IDENTITY, is used to concisely convey the semantics of an information module.

  Object definitions describe managed objects. An ASN.1 macro, OBJECT-TYPE, is used to concisely convey the syntax and semantics of a managed object.

  Notification definitions (also known as "traps") are used when describing unsolicited transmissions of management information. An ASN.1 macro, NOTIFICATION-TYPE, concisely conveys the syntax and semantics of a notification.

- **Management Information Base (MIB)**

  A management information base (MIB) stems from the OSI/ISO Network management model and is a type of database used to manage the devices in a communications network. It comprises a collection of objects in a (virtual) database used to manage entities (such as routers and switches) in a network.

  Objects in the MIB are defined using a subset of Abstract Syntax Notation One (ASN.1) called "Structure of Management Information Version

2 (SMIv2)" RFC 2578. The software that performs the parsing is a MIB compiler.

The database is hierarchical (tree-structured) and entries are addressed through object identifiers. See Figure 3.1 [Mi07].

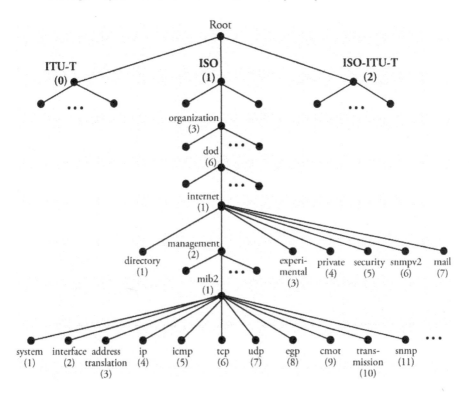

Figure 3.1: ASN.1 Object Identifier Organized Hierarchically

At the root of the object identifier hierarchy are three entries: ISO (International Standardization Organization), ITU-T (International Telecommunication Union – Telecommunication) standardization sector, and ISO-ITU-T, the joint branch of these two organizations. Figure 3.1 shows only part of the hierarchy. Under the ISO entry are other branches. For example, the organization (3) branch is labeled sequentially from the root as 1.3. If we continue to follow the entries on this branch, we see a path over dod (6), Internet (1), management (2), mib-2 (1), and ip (4). This path is identified by (1.3.6.1.2.1.4) to indicate all the labeled numbers from the root to the ip (4) entry. Besides that entry, MIB module represents a number of network interfaces and well-known Internet protocols at the bottom of this tree. This path clearly shows all the standards of "IP" associated with the "MIB-2" computer networking "management."

Internet documentation RFCs discuss MIBs, notably RFC 1155, "Structure and Identification of Management Information for TCP/IP based internets," and its two companions, RFC 1213, "Management Information Base for Network Management of TCP/IP-based internets," and RFC 1157, "A Simple Network Management Protocol."

The most basic elements of a network management model are graphically represented within the basic architecture of network management in Figure 3.2.

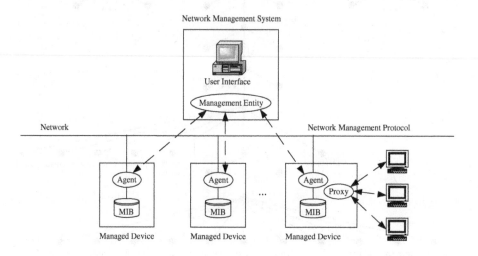

Figure 3.2: The Typical Network Management Architecture

Interactions between NMSs and managed devices can be any of four different types of commands: *read, write, traverse,* and *trap.*

- **Read**: To monitor managed devices, NMSs read variables maintained by the devices.

- **Write**: To control managed devices, NMSs write variables stored within the managed devices.

- **Traverse**: NMSs use these operations to determine which variables a managed device supports and to sequentially gather information from variable tables (such as IP routing tables) in managed devices.

- **Trap**: Managed devices use traps to asynchronously report certain events to NMSs.

3.2   Network Management Architectures

Most network management architectures use the same basic structure and set of relationships. End stations (managed devices), such as computer systems and other network devices, run software that enables them to send alerts when they recognize problems (for example, when one or more user-determined thresholds are exceeded). Upon receiving these alerts, management entities are programmed to react by executing one, several, or a group of actions, including operator notification, event logging, system shutdown, and automatic attempts at system repair. Management entities can also poll end stations to check the values of certain variables. Polling can be automatic or user-initiated, but agents in the managed devices respond to all polls. Agents are software modules that first compile information about the managed devices in which they reside, then store this information in a management database, and finally provide it (proactively or reactively) to management entities within network management systems (NMSs) via a network management protocol. Well-known network management protocols include the Simple Network Management Protocol (SNMP) and Common Management Information Protocol (CMIP). Management proxies are entities that provide management information on behalf of other entities. Figure 3.2 also depicts a typical network management architecture.

Networks, in essence, can be broadly classified as telecommunications networks and IP networks. Accordingly, current network management solutions have followed two general technical directions: ITU-T's Telecommunication Management Network (TMN) for telecommunications networks and IETF's Simple Network Management Protocol (SNMP) for IP networks. These two approaches adopt different standards, protocols, and implementations.

For the network management of telecommunications networks, it is derived from ITU M.3000 recommendation series building on open systems interconnection standards (OSI) and is known as Telecommunication Management Network (TMN). TMN is designed for public networks and is geared to two important objectives:

- functionality in a multi-vendor environment;

- optimization of network functionality.

And for IP networks, it is supported by IETF and based on Simple Network Management Protocol (SNMP), which has become the de facto standard in the management fields of IP networks. These two general models have thus adopted different standards and implementation methods, and are also designed for different network architectures [LS05]. The SNMP-based network management is chiefly for the handling of equipment in private data and networks, and to some extent for access access equipment.

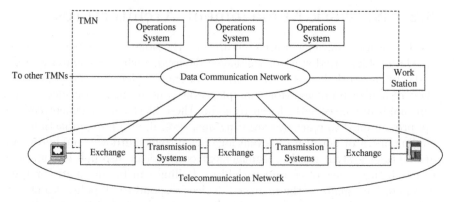

Note: The TMN boundary may extend to and manage customer/user services and equipment.

Figure 3.3: General Relationship of a TMN to a Telecommunication Network

3.2.1 TMN Management Architecture

Operation and maintenance is the classical term for control and supervision of telecommunications networks. However, the significant development of these activities in recent years has led to increased use of the term "network management". The purpose of network management for communication networks is two-fold:

- to enable the telecommunication network to provide customers with the services they demand, which is to create the greatest possible customer satisfaction,

- and to enable the operator to have these services provided at the lowest possible cost.

TMN has been widely adopted to manage telecommunications networks, ranging from transportation backbones to access networks. The TMN provides a structured framework for enabling interconnectivity and communication across heterogeneous operating systems and telecommunications networks. The TMN is defined in ITU M.3000 recommendation series, which cover a set of standards including common management information protocol (CMIP), guideline for definition of managed objects (GDMO), and abstract syntax notation one (ASN.1).

Recommendation M.3010 defines the general TMN management concepts and introduces several management architectures at different levels of abstraction:

- A functional architecture, which describes a number of management functions.

- A physical architecture, which defines how these management functions may be implemented into physical equipment.

- An information architecture, which describes concepts that have been adopted from OSI management.

- A logical layered architecture (LLA), which includes one of the best ideas of TMN: a model that shows how management can be structured according to different responsibilities.

Figure 3.3 presents the general relationship of a TMN to a telecommunication network.

Functional Architecture

Five different types of function blocks are defined by TMN's functional architecture. It is not necessary that all of these types are present in each possible TMN configuration. On the other hand, most TMN configurations will support multiple function blocks of the same type.

Figure 3.4 shows all five types of function blocks. In this figure, two types (OSF and MF) are completely drawn within the box labeled "TMN." This way of drawing indicates that these function blocks are completely specified by the TMN recommendations. The other three types (WSF, NEF, and QAF) are drawn at the edge of the box to indicate that only parts of these function blocks are specified by TMN. The following pages provide short descriptions of, plus the relation between, these five function blocks.

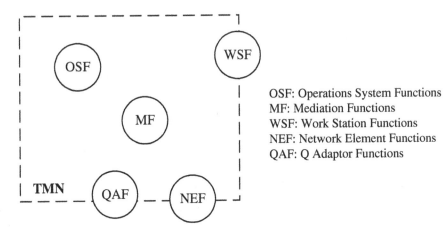

OSF: Operations System Functions
MF: Mediation Functions
WSF: Work Station Functions
NEF: Network Element Functions
QAF: Q Adaptor Functions

Figure 3.4: TMN Function Blocks

The TMN functional architecture introduces the concept of reference point to delineate function blocks. Five different classes of reference points are identified. Three of them (q, f, and x) are completely described by the TMN recommendations; the other classes (g and m) are located outside the TMN and only partially described.

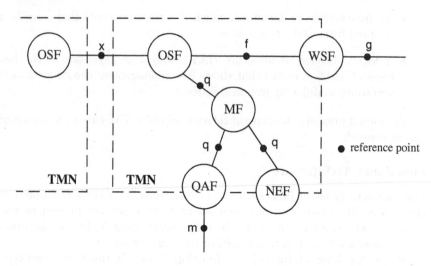

Figure 3.5: Example of Reference Points between Function Blocks

Figure 3.5 provides an example of reference points and function blocks. The picture shows for instance that the Mediation Function (MF) can be reached via q reference points and that the m reference point can be used to reach the Q Adaptor Function (QAF) from outside TMN.

Physical Architecture

TMN's physical architecture is defined at a lower abstraction level than TMN's functional architecture. See Figure 3.6.

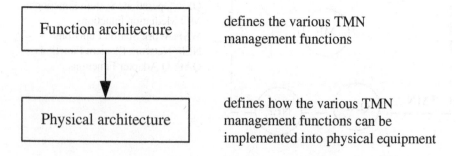

Figure 3.6: TMN Defined Multiple Related Architecture

The physical architecture shows how function blocks should be mapped upon building blocks (physical equipment) and reference points upon interfaces. In fact, the physical architecture defines how function blocks and reference points can be implemented. See Figure 3.7. It should be noted however that one function block may contain multiple functional components and one building block may implement multiple function blocks.

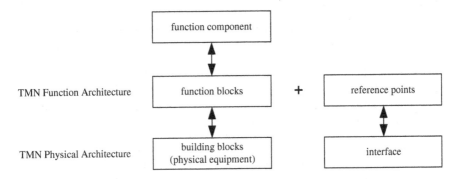

Figure 3.7: Relation between TMN Architectures

Information Architecture

TMN's information architecture uses an object oriented approach and is based on OSI's Management Information Model [ISO93]. According to this model, the management view of a managed object is visible at the managed object boundary. See Figure 3.8. At this boundary, the management view is described in terms of:

- Attributes, which are the properties or characteristics of the object.

- Operations, which are performed upon the object.

- Behavior, which is exhibited in response to operations.

- Notifications, which are emitted by the object.

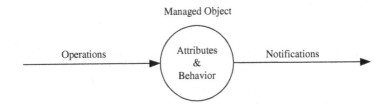

Figure 3.8: A Managed Object

The managed objects reside within managed systems, which include agent functions to communicate with the manager. TMN uses the same manager-agent concept as OSI.

Logical Layered Architecture

To deal with the complexity of management, in the framework of TMN, the following logical layers are defined:

- Network Elements (NE): are involved with the management functionality that network element itself supports, independent of any management system. Network element layer is actually tremendously important to the effectiveness of management systems.

- Element Management Layer (EML): involves managing the individual devices in the network and keeping them running. This includes functions to view and change a network element's configuration, to monitor alarm messages emitted from elements in the network, and to instruct network elements to run self-tests.

- Network Management Layer (NML): involves managing relationships and dependencies between network elements, generally required to maintain end-to-end connectivity of the network. It is concerned with keeping the network running as a whole. In contrast, although element management enables the management of every element in the network, it does not cover functions that deal with ensuring overall network integrity. It is possible, for example, to have a network with individual element configurations that are perfectly valid but that do not match up properly. As a consequence, the network does not work as intended. For example, to configure a static path across the network, each element along the path must be configured properly. Otherwise, the path is broken and data cannot reach its destination. Likewise, timer values need to be tuned to avoid excessive timeouts and retransmissions. Monitoring tasks at the network management layer involves ensuring that data flows across the network and reaches its destination with acceptable throughput and delay. Policies that control the kinds of calls to admit at any given entry point into the network need to be coordinated across the network to be effective.

  These kinds of tasks are addressed at the network management layer. It takes into account the networking context of the individual devices and involves managing the end-to-end aspects of the network. It offers the concept of a forest, as opposed to individual trees. An example of a network management task is the management of a network connection as a whole, for instance, setting it up and monitoring it. As mentioned earlier, this involves managing multiple devices in a concerted fashion. Such management includes not only managing how devices are configured individually, but also ensuring that their configurations are coordinated in certain ways and monitoring for cross-network connectivity, instead of and in addition to simply ensuring that individual elements are up and running. The network management layer makes use of functionality provided by the element management layer, providing additional functions on top.

- Service Management Layer (SML): is concerned with managing the services that the network provides and ensuring that those services are running smoothly and functioning as intended. For example, when a customer orders a service, the service needs to be turned up. This might be required for a new employee in an enterprise who needs phone service. Turning up phone service might, in turn, result in a number of operations that need to be carried out across the network so that the service is activated: A phone number must be allocated. The company directory must be updated. Voicemail servers and IP PBXs need to be made aware of the new extension. Later, the user might call the service help desk and complain that the service is not working properly. Problems could include poor voice quality and calls that disconnected unexpectedly. Troubleshooting the service is required to identify the root cause of the problem and solve it. These are all examples of typical tasks in managing a service. These tasks build on functionality that is provided by the network management layer underneath and provide additional value on top, applying them to the context of managing a service.

- Business Management Layer (BML): deals with managing the business associated with providing services and all the required support functions. This includes topics as diverse as billing and invoicing, helpdesk management, business forecasting, and many more.

The functionalities of TMN are defined to cover 5 major areas: configuration management, fault management, performance management, accounting management, and security management. Such kinds of two-dimensional partition provides a well-structured framework for developing network management system (see Figure 3.9). That is why the "TMN" concept is so popular, and can be seen in various implementations of network management technologies. The CMIP- and CORBA-based management solutions are two typical representatives of TMN applications.

3.2.2 Internet-Based Management Architecture

Current IP networks are often managed via Simple Network Management Protocol (SNMP), which is pushed by IETF as a specification, initially presented for the Internet. So far, there have been several versions of SNMP. The

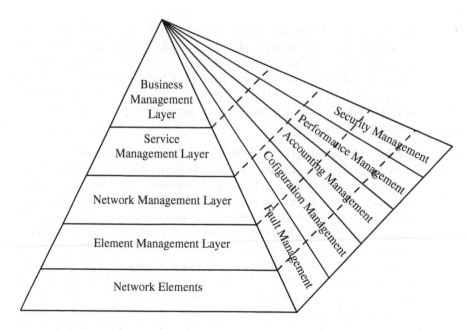

Figure 3.9: Management Layer Model and Function Areas

common ones are SNMPv1, SNMPv2, and SNMPv3. The SNMP is an application layer protocol and uses User Datagram Protocol (UDP) to exchange management information between management entities.

SNMP Model for Network Management

SNMP is part of a larger architecture, called the Internet Network Management Framework (NMF). The Internet Standard Management Framework encompasses all of the technologies that comprise the TCP/IP network management solution. The SNMP Framework consists of a number of architectural components that define how management information is structured, how it is stored, and how it is exchanged using the SNMP protocol. The Framework also describes how the different components fit together, how SNMP is to be implemented in network devices, and how the devices interact.

SNMP Framework Components

As we will explore in more detail later, the Internet Standard Management Framework is entirely information-oriented. It includes the following primary components (see Figure 3.10):

- Structure of Management Information (SMI)

- Management Information Bases (MIBs)

- Simple Network Management Protocol (SNMP)

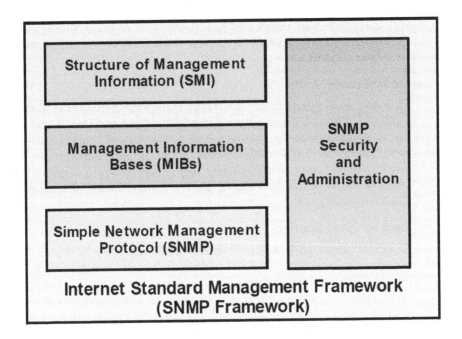

Figure 3.10:   Components of the TCP/IP Internet Standard Management Framework

- Security and Administration

TCP/IP SNMP network management system comprises some basic components:

- SNMP Device Types

  As we saw in the preceding high-level overview topic, the overall idea behind SNMP is to allow the information needed for network management to be exchanged using TCP/IP. More specifically, the protocol allows a network administrator to make use of a special network device that interacts with other network devices to collect information from them and modify how they operate. In the simplest sense, then, two different basic types of hardware devices are defined:

  - Managed Nodes: Regular nodes on a network that have been equipped with software to allow them to be managed using SNMP. These are, generally speaking, conventional TCP/IP devices; they are also sometimes called managed devices.

  - Network Management Station (NMS): A designated network device that runs special software to allow it to manage the regular managed nodes mentioned just above. One or more NMSes must be present on the network, as these devices are the ones that really "run" SNMP.

- SNMP Entities

Each device that participates in network management using SNMP runs a piece of software, generically called an SNMP entity. The SNMP entity is responsible for implementing all of the various functions of the SNMP protocol. Each entity consists of two primary software components. Which components comprise the SNMP entity on a device depends of course on whether the device is a managed node or a network management station.

- Managed Node Entities

An SNMP managed node can be pretty much any network device that can communicate using TCP/IP, as long as it is programmed with the proper SNMP entity software. SNMP is designed to allow regular hosts to be managed, as well as intelligent network interconnection devices such as routers, bridges, hubs, and switches. Other "unconventional" devices can likewise be managed, as long as they connect to a TCP/IP internetwork: printers, scanners, consumer electronic devices, even special medical devices and more.

The SNMP entity on a managed node consists of the following software elements and constructs:

  - SNMP Agent: A software program that implements the SNMP protocol and allows a managed node to provide information to an NMS and accept instructions from it.

  - SNMP Management Information Base (MIB): Defines the types of information stored about the node that can be collected and used to control the managed node. Information exchanged using SNMP takes the form of objects from the MIB.

- Network Management Station Entities

On a larger network, a network management station may be a separate, high-powered TCP/IP computer dedicated to network management. However, it is really software that makes a device into an NMS, so the NMS may not be a separate hardware device. It may act as an NMS and also perform other functions on the network.

The SNMP entity on a network management station consists of:

  - SNMP Manager: A software program that implements the SNMP protocol, allowing the NMS to collect information from managed nodes and to send instructions to them.

     – SNMP Applications: One or more software applications that allow
a human network administrator to use SNMP to manage a network.

SNMP consists of a small number of network management stations (NMSs)
that interact with regular TCP/IP devices that are called managed nodes.
The SNMP manager on the NMS and the SNMP agents on the managed
nodes implement the SNMP protocol and allows network management information to be exchanged. SNMP applications run on the NMS and provide the
interface to the human administrator, and allow information to be collected
from the MIBs at each SNMP agent, as shown in Figure 3.11 [Koz05].

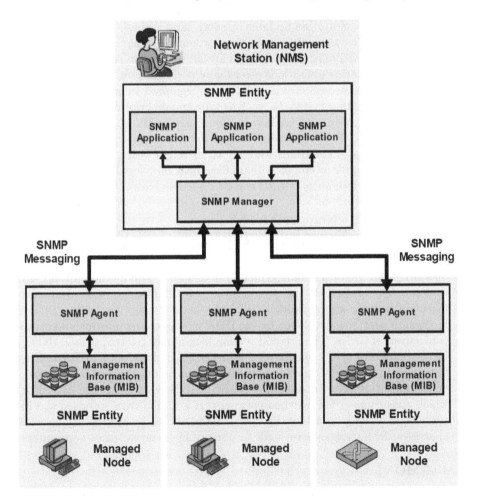

Figure 3.11: SNMP Operational Model

In typical SNMP usage, there are a number of systems to be managed, and one or more systems managing them. A software component called an agent runs on each managed system and reports information via SNMP to the managing systems.

Essentially, SNMP agents expose management data on the managed systems as variables (such as "free memory," "system name," "number of running processes," "default route"). But the protocol also permits active management tasks, such as modifying and applying a new configuration. The managing system can retrieve the information through the GET, GETNEXT, and GETBULK protocol operations or the agent will send data without being asked using TRAP or INFORM protocol operations. Management systems can also send configuration updates or controlling requests through the SET protocol operation to actively manage a system. Configuration and control operations are used only when changes are needed to the network infrastructure. The monitoring operations are usually performed on a regular basis.

The variables accessible via SNMP are organized in hierarchies. These hierarchies, and other metadata (such as type and description of the variable), are described by Management Information Bases (MIBs).

Although SNMP is only a lightweight implementation for network management, additional standards were added in recent years, such as SNMPv3 and RMON in order to enhance its management functionalities, especially in security and performance. SNMP will be disscussed in detail in Section 3.3.

RMON: Remote Network Monitoring

The most important addition to the basic set of SNMP standards is the RMON (Remote Network MONitoring) standard, RFC 1271. RMON is a major step forward in internetwork management. It defines a remote-monitoring MIB that supplements MIB-II and provides the network manager with vital information about the internetwork.

RMON MIB was developed by the IETF to support monitoring and protocol analysis of LANs. The original version (sometimes referred to as RMON1) focused on OSI Layer 1 and Layer 2 information in Ethernet and Token Ring networks. It has been extended by RMON2 which adds support for Network- and Application-layer monitoring and by SMON (Oracle System MONitor) which adds support for switched networks. It is an industry standard specification that provides much of the functionality offered by proprietary network analyzers. RMON agents are built into many high-end switches and routers.

1. RMON1

With the RMON1 MIB, network managers can collect information from remote network segments for the purposes of troubleshooting and performance monitoring. The RMON1 MIB provides:

- Current and historical traffic statistics for a network segment, for a specific host on a segment, and between hosts (matrix).

- A versatile alarm and event mechanism for setting thresholds and notifying the network manager of changes in network behavior.

- A powerful, flexible filter and packet capture facility that can be used to deliver a complete, distributed protocol analyzer.

The Figure 3.12 shows a listing of the RMON1 groups and where RMON fits within the International Standards Organization (ISO) and IETF standards.

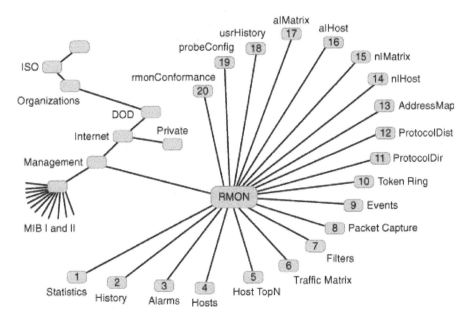

Figure 3.12: RMON MIB Tree Diagram

An RMON implementation typically operates in a client/server model. Monitoring devices (commonly called "probes" in this context) contain RMON software agents that collect information and analyze packets. These probes act as servers and the Network Management applications that communicate with them act as clients. While both agent configuration and data collection use SNMP, RMON is designed to operate differently than other SNMP-based systems:

Probes have more responsibility for data collection and processing, which reduces SNMP traffic and the processing load of the clients.

Information is only transmitted to the management application when required, instead of continuous polling.

In short, RMON is designed for "flow-based" monitoring, while SNMP is often used for "device-based" management. RMON is similar to other flow-based monitoring technologies such as NetFlow and SFlow because the data collected deals mainly with traffic patterns rather than the status of individual devices. One disadvantage of this system is that remote devices shoulder more of the management burden and require more resources to do so. Some devices balance this trade-off by implementing only a subset of the RMON MIB groups (see below). A minimal RMON agent implementation could support only statistics, history, alarm, and event.

The RMON1 MIB consists of ten groups:

1. Statistics: real-time LAN statistics, e.g., utilization, collisions, CRC errors.

2. History: history of selected statistics.

3. Alarm: definitions for RMON SNMP traps to be sent when statistics exceed defined thresholds.

4. Hosts: host specific LAN statistics, e.g., bytes sent/received, frames sent/received.

5. Hosts top N: record of N most active connections over a given time period.

6. Matrix: the sent-received traffic matrix between systems.

7. Filter: defines packet data patterns of interest, e.g., MAC address or TCP port.

8. Capture: collect and forward packets matching the Filter.

9. Event: send alerts (SNMP traps) for the Alarm group.

10. Token Ring: extensions specific to Token Ring.

Capabilities of RMON1

- Without leaving the office, a network manager can watch the traffic on a LAN segment, whether that segment is physically located around the corner or around the world. Armed with that traffic knowledge, the network manager can identify trends, bottlenecks, and hotspots. When a problem arises, RMON1 also includes a powerful protocol analyzer so the network manager has distributed troubleshooting tools immediately at hand. Since the RMON1 device is permanently attached to the network segment, it's already collecting data about the remote LAN and ready to transmit it to a central network management station whenever required. All this network monitoring and troubleshooting can be done without spending the time and travel required to send expensive network experts with "lug-able" protocol analyzers to the remote site.

- Deploying network management staff resources more efficiently means that one expert at a central site can be working on several problems by getting information from several probes at remote sites. Alternatively, several experts with different specialties can be focused on a single segment by getting information from a single probe.

- Network managers desperately need tools that can leverage their resources and increase their scope of control. RMON1 does just that. A recent study by McConnell Consulting found that by using RMON1 distributed LAN management and remote monitoring techniques, a network management team can support as many as two-and-a-half times the users and segments without adding staff. Using RMON1 and the network to bring the problem to the expert is far more cost-efficient than dispatching someone to the remote site with a portable protocol analyzer.

2. RMON2

The RMON2 Working Group began their efforts in July 1994. As with the RMON1 standard, the approach is to carve out a set of deliverables that bring clear benefits to the network manager, that are implementable by multiple vendors, and that will lead to successful interoperability between independently developed solutions.

With those broad goals in mind, the top priority defined by the RMON2 Working Group is to go up the protocol stack and provide statistics on network- and application-layer traffic. By monitoring at the higher protocol layers, RMON2 provides the information that network managers need to see beyond the segment and get an internetwork or enterprise view of network traffic.

The RMON2 MIB adds ten more groups:

1. Protocol Directory: list of protocols the probe can monitor.

2. Protocol Distribution: traffic statistics for each protocol.

3. Address Map: maps network-layer (IP) to MAC-layer addresses.

4. Network-Layer Host: layer 3 traffic statistics, per each host.

5. Network-Layer Matrix: layer 3 traffic statistics, per source/destination pairs of hosts

6. Application-Layer Host: traffic statistics by application protocol, per host.

7. Application-Layer Matrix: traffic statistics by application protocol, per source/destination pairs of hosts.

8. User History: periodic samples of user-specified variables.

9. Probe Configuration: remote config of probes.

10. RMON Conformance: requirements for RMON2 MIB conformance.

Capabilities of RMON2

The most visible and most beneficial capability in RMON2 is monitoring above the MAC layer, which supports protocol distribution and provides a view of the whole network rather than a single segment. Although the exact contents of RMON2 may change during the standard development process, the capabilities expected to be delivered by RMON2 include:

- Higher Layer Statistics. Traffic statistics, host, matrix, and matrix topN tables at the network layer, and the application layer. By monitoring these statistics, the network manager can see which clients are talking to which servers, so systems can be placed at the correct location on the correct segment for optimized traffic flow. Figure 3.13 denotes the work layers of RMON1 and RMON2.

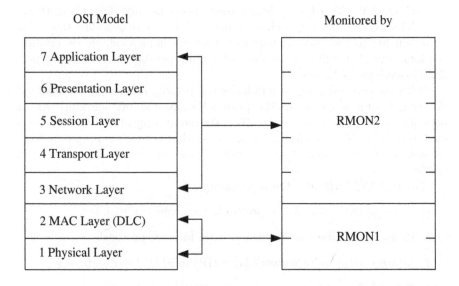

Figure 3.13: The Work Layers of RMON1 and RMON2

- Address Translation. Binding between MAC-layer addresses and network-layer addresses, which are much easier to read and remember. Address translation not only helps the network manager, it supports the SNMP management platform and will lead to improved topology maps. This feature also adds duplicate IP address detection, solving an often elusive problem that wreaks havoc with network routers and Virtual LANs.

- User-Defined History. With this new feature, the network manager can configure history studies of any counter in the system, such as a specific history on a particular file server or a router-to-router connection. In the RMON1 standard, historical data are collected only on a predefined set of statistics.

- Improved Filtering. Additional filters are required to support the higher-layer protocol capabilities of RMON2. This improved filtering allows the user to configure more flexible and efficient filters, especially relating to the higher-layer protocols.

- Probe Configuration. With RMON2, one vendor's RMON application will be able to remotely configure another vendor's RMON probe. Currently, each vendor provides a proprietary means of setting up and controlling their probes. The probe configuration specification is based on the Aspen MIB which was jointly developed by AXON and Hewlett-Packard. The Aspen MIB provides probe device configuration, trap administration, and control of the probe's out-of-band serial port.

3.2.3 Comparison of TMN- and Internet-Based Management Architecture

The comparisons between TMN and Internet (SNMP) based management models are as follows [LS05]:

- Complexity

  TMN: Feature-rich modeling of managed objects described in GDMO. However, the data modeling and abstracting are very complex because of TMS's fine-grained definition for interface and object.

  SNMP: Simplified design and architecture. In addition, the variables in SNMP can be easily programmed. It is simple and easy to use.

- Functionality

  TMN gives a general framework for network management, and major functional areas that have been widely accepted in industry. Other security features are also included, such as access control and security logging. The use of data communication networks (DCN) for internal communication makes it physically secure.

  SNMP follows TMN's framework for management functionalities. But, SNMP agents can only collect information from devices, lacking the ability of analyzing. The openness and IP-oriented nature of SNMP makes it not secure as TMN-based protocols, such as CMIP, which defines management services exchanged between peer entities in TMN.

- Multi-vendor

  Multi-vendor support is achieved at network management layer by implementing an interface between EMS and NMS. NMS can exchange
  events via its northbound interface with different EMSs that have provided a southbound interface. However, practically, it is difficult and
  expensive to implement NMS-EMS interface because of the complexity
  of TMN.

  Multi-vendor support can be offered by retrieving objects from public
  MIBs (e.g., SNMPv1) that reside in the managed devices of different
  vendors. While, for private MIBs, the interface for specific vendor has
  to be developed.

- Communication

  The communication between NE/NMS and EMS requires the special
  OSI protocol stacks, which are rarely supported by common LAN or
  WAN. SNMP is initially designed for IP technology and uses UDP to
  carry management data. It can easily run on nearly any network because
  of the popularity of TCP/IP. However, SNMP is connectionless with
  lower overhead, and thus can't guarantee the deliver of messages.

- Implementation

  Taking CMIP and CORBA architecture as an example, the development
  of the core components in TMN has to rely on many third-party software
  packages. The implementation and running of TMN systems have higher
  requirements to networks and operation systems.

  The development of SNMP interface is relatively simple because of the
  simplicity of the standards and availability of TCP/IP protocol used.
  The cost of implementing SNMP network management is much lower,
  compared to the development of TMN-based architecture

In general, SNMP model is simple, cost-effective, and open in standards.
The simplicity and ease of implementation of SNMP is why it is the most
popular protocol for managing networks. In contrast, the CMIP- or CORBA-
based TMN models are initially proposed for the management of telecommunications networks, and concentrate on reliability and stability of networks.
Because of the incurred complexity, it requires more resources to develop and
run. Therefore, it is most suitable for some mission critical applications, such
as the management of transportation backbones.

3.3 Evolution of Network Management Protocols

With the expansion of networks, the evolution of network architectures, and
the increasing requirements for network management, the network manage-

ment protocols also evolve consequently.

A typical management network system will make use of the management operation services to monitor network elements. Management agents found on network elements will make use of the management notification services to send notifications or alarms to the network management system.

Network management protocols are used to define how network management information is exchanged between network management services and management agents. Some popular network management protocols are discussed as follows:

3.3.1 Common Management Information Protocol (CMIP)

Common Management Information Protocol (CMIP) is an OSI-based network management protocol. It provides an implementation for the services defined by CMIS (the Common Management Information Service), allowing communication between network management applications and management agents. CMIS/CMIP emerged out of the ISO/OSI network management model and is defined by the ITU-T X.700 series of recommendations, its more popular correspondent designed by the IETF being SNMP.

CMIP adopts an ISO reliable connection-oriented transport mechanism and has built up security that supports access control, authorization, and security logs. The management information is exchanged between the network management application and management agents through managed objects.

CMIP models management information in terms of managed objects and allows both modification and performing actions on managed objects. Managed objects are described using GDMO (the Guidelines for the Definition of Managed Objects) and are identified by a distinguished name (DN), similar in concept to the X.500 directory.

CMIP also provides good security (support authorization, access control, and security logs) and flexible reporting of unusual network conditions.

The management functionality implemented by CMIP is described under CMIS services.

CMIS is a service that may be employed by network elements for network management. It defines the service interface that is implemented by the Common management information protocol (CMIP). CMIS is part of the Open Systems Interconnection (OSI) body of network standards.

Note the term CMIP is sometimes used erroneously when CMIS is intended. CMIS/CMIP is most often used in telecommunication applications, in other areas SNMP has become more popular.

The following services are made available by the Common Management Information Service Element (CMISE) to allow management of network elements:

- Management operation services

  - M-CREATE: Create an instance of a managed object

- M-DELETE: Delete an instance of a managed object

- M-GET: Request managed object attributes (for one object or a set of objects)

- M-CANCEL-GET: Cancel an outstanding GET request

- M-SET: Set managed object attributes

- M-ACTION: Request an action to be performed on a managed object

- Management notification services

  - M-EVENT-REPORT: Send events occurring on managed objects

- Management association services

  To transfer management information between open systems using CMIS/CMIP, peer connections, i.e., associations, must be established. This requires the establishment of an Application layer association, a Session layer connection, a Transport layer connection, and, depending on supporting communications technology, Network layer, and Link layer connections.

  CMIS initially defined management association services but it was later decided these services could be provided by ACSE and these services were removed. Below is a list of these services which were subsequently removed from ISO 9595:

  - M-INITIALIZE - Creates an association with (i.e. connects to) another CMISE

  - M-TERMINATE - Terminates an established connection

  - M-ABORT - Terminates the association in the case of an abnormal connection termination

The CMIP over the OSI Management Architecture

CMIP is widely used in the telecommunication domain and telecommunication devices typically support CMIP. The International Telecommunication Union (ITU) endorses CMIP as the protocol for the management of devices in the Telecommunication Management Network (TMN) standard.

CMIP is an ISO development, and it is designed to operate in the OSI environment. It is considerably more complex than its SNMP counterpart. Figure 3.14 illustrates the OSI Management architecture, which uses CMIP to access managed information.

CMIP is part of the ITU-T X.700 OSI series of recommendations of the ITU. CMIP was developed and funded by government and corporations to replace and make up for the deficiencies of SNMP, thus improving the capabilities of network management systems.

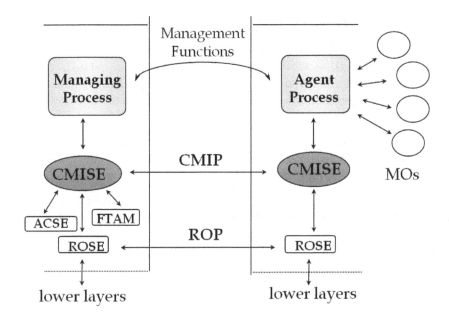

Figure 3.14: CMIP on the OSI Management Architecture

In CMIP, an agent maintains a management information tree (MIT) as a database; it models platforms and devices using managed objects (MOs). These may represent LANs, ports, and interfaces. CMIP is used by a platform to change, create, retrieve, or delete MOs in the MIT. It can invoke actions or receive event notifications.

Object-oriented system concepts that are applied to the CMIP objects include containment, inheritance, and allomorphism. Containment refers to the characteristic of objects being a repository of other objects and/or attributes. A high-level object for a communication switch, for example, can contain several racks of equipment, each of which, in turn, can contain several slots for printed circuit boards. Here one might use the ITU-T M.3100 base class for a circuit pack to define the general features of modules within a communication switch. Object classes can then, in turn, be defined to represent the specific modules. Items including line interface cards, switching elements, and processors can be derived from the basic circuit pack definition. Each of these objects exhibits the behavior, actions, and attributes of both the derived classes and the base class. Allomorphism is a concept coined by the CMIP standards bodies to refer to the ability to interact with modules through a base set of interfaces, only to have the resulting behaviors coupled to the complete class definition. Disabling a power supply, for instance, may exhibit significantly different behavior than disabling a switching component [AP93][AP93]. With CMIP and other OSI management schemes, there are three types of relationships between managed objects:

- Inheritance Tree. This defines the managed object class super and sub-classes, much as C++ base and derived classes are related. When a class is inherited from a superclass, it possesses all the characteristics of the superclass, with additional class-specific extensions (additional attributes, behaviors, and actions).

- Containment Tree. This defines which managed objects are contained in other managed objects. As an example, a subnetwork can contain several managed elements (ME).

- Naming Tree. This defines the way in which individual objects are referenced within the constraints of the management architecture.

CMIP (i.e., OSI management communications) communications are very different from those found in SNMP. These communications are embedded in the OSI application environment and they rely on conventional OSI peer layers for support. They use connection-oriented transport where SNMP uses the datagram (connectionless). In most cases, these communications are acknowledged.

The CMIP Over TCP/IP (CMOT) Management Architecture

The CMOT (CMIP Over TCP/IP) architecture is based on the OSI management framework and the models, services, and protocols developed by ISO for network management. The CMOT architecture demonstrates how the OSI management framework can be applied to a TCP/IP environment and used to manage objects in a TCP/IP network. The use of ISO protocols for the management of widely deployed TCP/IP networks will facilitate the ultimate migration from TCP/IP to ISO protocols. The concept of proxy management is introduced as a useful extension to the architecture. Proxy management provides the ability to manage network elements that either are not addressable by means of an Internet address or use a network management protocol other than CMIP. The CMOT architecture specifies all the essential components of a network management architecture. The OSI management framework and models are used as the foundation for network management. A protocol-dependent interpretation of the Internet SMI is used for defining management information. The Internet MIB provides an initial list of managed objects. Finally, a means is defined for using ISO management services and protocols on top of TCP/IP transport protocols. Management applications themselves are not included within the scope of the CMOT architecture. What is currently standardized in this architecture is the minimum required for building an interoperable multivendor network management system. Applications are explicitly left as a competitive issue for network developers and providers.

The objective of the CMOT protocol architecture is to map the OSI management protocol architecture into the TCP/IP environment. The model presented here follows the OSI model at the application layer, while using Internet protocols at the transport layer. Figure 3.15 denotes the CMOT Protocol Architecture. To guarantee reliable transport, CMOT systems establish Application layer connections prior to transmitting management information. CMOT's Application layer services are built on three OSI services: the Common Management Information Service Element (CMISE), the Remote Operation Service Element (ROSE), and the Association Control Service Element (ACSE). A Lightweight Presentation Protocol (LPP) provides Presentation layer services.

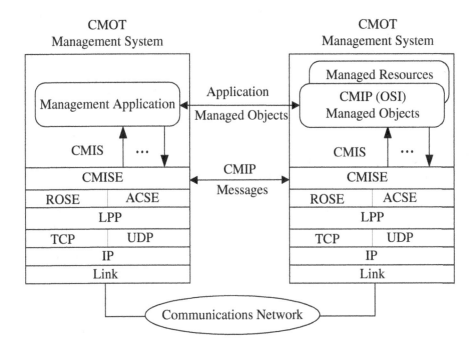

Figure 3.15: The CMOT Protocol Architecture

3.3.2 Simple Network Management Protocol (SNMP)

SNMP is an application layer protocol based on the Transmission Control Protocol/Internet Protocol (TCP/IP) protocol suite, which offers network management services for monitoring and control of network devices. SNMP enables network administrators to manage network performance, find and solve network problems, and plan for network growth. SNMP is a network management tool that allows network administrator to perform monitoring, control, and planning tasks on the network to be managed.

In the early days of the Internet, the Internet Activities Board recognized the need for a management framework by which to manage TCP/IP implementations. The framework consists of three components:

1. A conceptual framework that defines the rules for describing management information, known as the Structure of Management Information (SMI).

2. A virtual database containing information about the managed device known as the management Information Base (MIB).

3. A protocol for communication between a manager and an agent of a managed device, know as Simple Network Management Protocol (SNMP).

Essentially, the data that are handled by SNMP must follow the rules for objects in the MIB, which in turn are defined according to the SMI.

SNMP is an application layer protocol that is used to read and write variables in an agent's MIB.

SNMP is based on an asynchronous request-response protocol enhanced with trap-directed polling. The qualifier asynchronous refers to the fact that the protocol need not wait for a response before sending other messages. Trap-directed polling refers to the fact that a manager polls in response to a trap message being sent by an agent, which occurs when there is an exception or after some measure has reached a certain threshold value. SNMP operates in a connectionless manner with UDP being the preferred transport mode. An SNMP manager sends message to an agent via UDP destination port 161, while an agent sends trap messages to a manager via UDP destination port 162. The connectionless mode was chosen partly to simplify SNMP's implementation and because connectionless is usually the preferred mode for management applications that need to talk to many agents.

Figure 3.16 presents the detail operation of SNMP.

The normal SNMP manager operations are identified as in Table 3.1.

There are currently three versions of SNMP: SNMPv1, SNMPv2, and SNMPv3.

The modular design of SNMP is shown in the consistency of the architecture, structure, and framework of all three versions; this aids gradual evolution of protocol enhancements. Though SNMPv1 was effective and easy to implement, it had its problems and limitations. Enhancements to SNMPv1, resulted in a new SNMP version, SNMPv2, which also corrected the bugs and limitations in SNMPv1. However, these new enhancements did not address security deficiencies, such as privacy of data, masquerading, and unauthorized disclosure of data. Subsequently, SNMPv3 was then developed to address these security deficiencies: SNMPv3 added security features, such as access control, authentication, and encryption of management data. The

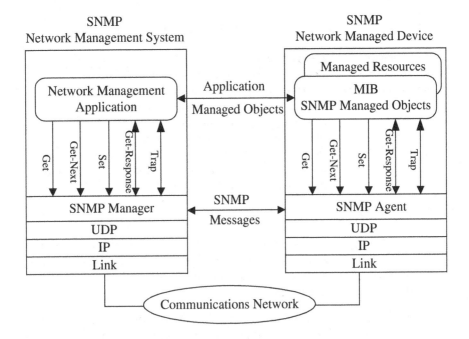

Figure 3.16: SNMP Protocol

SNMPv3 specifications were approved by the Internet Engineering Steering Group (IESG) as full Internet Standard in March 2002, and vendors have begun to support SNMPv3 in their products.

SNMPv1

SNMPv1 is the original Internet-Standard Network Management Framework, as described in RFCs 1155, 1157, and 1212. There are typically three communities in SNMPv1: read-only, read-write, and trap. It should be noted that while SNMPv1 is historical, it is still the primary SNMP implementation that many vendors support.

SNMPv1's security is based on communities, which are nothing more than passwords: plain-text strings that allow any SNMP-based application that knows the strings to gain access to a device's management information.

SNMPv1 Protocol Operations

SNMPv1 is a simple request/response protocol. The network-management system issues a request, and managed devices return responses. This behavior is implemented by using one of four protocol operations: Get, GetNext, Set, and Trap.

Table 3.1: SNMP Manager Operations

Operation	Description
get-request	Retrieve a value from a specific variable.
get-next-request	Retrieve a value from a variable within a table.
get-response	The reply to a get-request, get-next-request, and set-request sent by an NMS.
set-request	Store a value in a specific variable.
trap	An unsolicited message sent by an SNMP agent to an SNMP manager indicating that some event has occurred.

- The Get operation is used by the NMS to retrieve the value of one or more object instances from an agent. If the agent responding to the Get operation cannot provide values for all the object instances in a list, it does not provide any values.

- The GetNext operation is used by the NMS to retrieve the value of the next object instance in a table or a list within an agent.

- The Set operation is used by the NMS to set the values of object instances within an agent.

- The Trap operation is used by agents to asynchronously inform the NMS of a significant event.

The SNMPv1 Framework describes the encapsulation of SNMPv1 PDUs in SNMP messages between SNMP entities and distinguishes between application entities and protocol entities. In SNMPv3, these are renamed applications and engines, respectively.

The SNMPv1 Framework also introduces the concept of an authentication service supporting one or more authentication schemes. In SNMPv3, the concept of an authentication service is expanded to include other services, such as privacy.

Finally, the SNMPv1 Framework introduces access control based on a concept called an SNMP MIB view. The SNMPv3 Framework specifies a fundamentally similar concept called view-based access control.

However, while the SNMPv1 Framework anticipated the definition of multiple authentication schemes, it did not define any such schemes other than a trivial authentication scheme based on community strings. This was a known fundamental weakness in the SNMPv1 Framework. However, at that time, it was thought that the definition of commercial grade security might be contentious in its design and difficult to get approved because "security" means many different things to different people. To that end, and because some users do not require strong authentication, the SNMPv1 structured an authentication service as a separate block to be defined "later." The SNMPv3 Framework provides an architecture for use within that block, as well as a definition for its subsystems.

SNMPv2

SNMPv2 is derived from the SNMPv1 framework. It is described in STD 58, RFCs 2578, 2579, 2380, and STD 62, RFCs 3416, 3417, and 3418. SNMPv2 has no message definition.

SNMPv2 Protocol Operations

The Get, GetNext, and Set operations used in SNMPv1 are exactly the same as those used in SNMPv2. However, SNMPv2 adds and enhances some protocol operations. The SNMPv2 Trap operation, for example, serves the same function as that used in SNMPv1, but it uses a different message format and is designed to replace the SNMPv1 Trap.

SNMPv2 also defines two new protocol operations: GetBulk and Inform.

- The GetBulk operation is used by the NMS to efficiently retrieve large blocks of data, such as multiple rows in a table. GetBulk fills a response message with as much of the requested data as will fit. In SNMPv2, if the agent responding to GetBulk operations cannot provide values for all the variables in a list, it provides partial results.

- The Inform operation allows one NMS to send trap information to another NMS and to then receive a response.

SNMPv2 provides several advantages over SNMPv1:

- Expanded data types: 64-bit counter

- Improved efficiency and performance: get-bulk operator

- Confirmed event notification: inform operator

- Richer error handling: errors and exceptions

- Improved sets: especially row creation and deletion

- Fine-tuned data definition language

  However, the SNMPv2 framework, as described in RFCs 1902 – 1907, is incomplete in that it does not meet the original design goals of the SNMPv2 project. The unmet goals include provision of security and administration delivering so-called "commercial grade" security with

  – authentication: origin identification, message integrity, and some aspects of replay protection,

- privacy: confidentiality,

- authorization and access control, and

- suitable remote configuration and administration capabilities for these features.

SNMPv2c (the Community-based SNMP version 2) is an experimental SNMP framework which supplements the SNMPv2 Framework, as described in RFC 1901. It adds the SNMPv2c message format, which is similar to the SNMPv1 message format.

SNMPv3

SNMPv3 is described in STD 62, RFCs 3412, 3414, and 3417. Coexistence issues relating to SNMPv1, SNMPv2c, and SNMPv3 can be found in RFC 3416. The new features of SNMPv3 (in addition to those of SNMPv2 listed above) include:

- Security

  - authentication and privacy
  - authorization and access control

- Administrative Framework

  - naming of entities
  - people and policies
  - usernames and key management
  - notification destinations
  - proxy relationships
  - remotely configurable via SNMP operations

SNMP version 3 (SNMPv3) is the latest version of SNMP. Its main contribution to network management is security. It adds support for strong authentication and private communication between managed entities. In 2002, it finally made the transition from draft standard to full standard. While it is good news that SNMPv3 is a full standard, vendors are notoriously slow at adopting new versions of a protocol. While SNMPv1 has been transitioned to historical, the vast majority of vendor implementations of SNMP are SNMPv1 implementations. Some large infrastructure vendors like Cisco have supported SNMPv3 for quite some time, and we will undoubtedly begin to see more vendors move to SNMPv3 as customers insist on more secure means of managing networks.

SNMPv3 is an interoperable standards-based protocol for network management. SNMPv3 provides secure access to devices by a combination of authenticating and encrypting packets over the network. The security features provided in SNMPv3 are:

- Message integrity: Ensuring that a packet has not been tampered with in-transit.

- Authentication: Determining the message is from a valid source.

- Encryption: Scrambling the contents of a packet prevent it from being seen by an unauthorized source.

SNMPv3 provides for both security models and security levels. A security model is an authentication strategy that is set up for a user and the group in which the user resides. A security level is the permitted level of security within a security model. A combination of a security model and a security level will determine which security mechanism is employed when handling an SNMP packet. Three security models are available: SNMPv1, SNMPv2c, and SNMPv3. Figure 3.17 denotes the SNMPv3 entity.

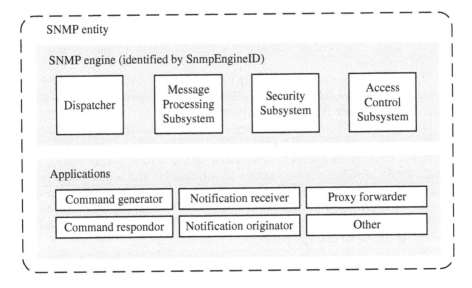

Figure 3.17: SNMPv3 Entity

Recent Advances in SNMP

To improve the ability of SNMP for configuring networks and devices, RFC 3512 (2003) offers guidance in the effective use of SNMP for configuration management. This information is relevant to vendors that build network elements, management application developers, and those that acquire and deploy this technology in their networks.

To efficiently enrich the ability of processing network management information, RFC 3781 (2004) defines an SMIng (Structure of Management Information, Next Generation) language extension that specifies the mapping of SMIng definitions of identities, classes, and their attributes and events to dedicated definitions of nodes, scalar objects, tables and columnar objects, and notifications for application in the SNMP management framework.

To enhance the security of SNMP, RFC 3826 (2004) describes a symmetric encryption protocol that supplements the protocols described in the User-based Security Model (USM), which is a Security Subsystem for SNMPv3 in the SNMP Architecture. The symmetric encryption protocol is based on the Advanced Encryption Standard (AES) cipher algorithm used in Cipher FeedBack Mode (CFB), with a key size of 128 bits.

To enlarge the management ability for increasing heterogeneity communication devices, network management has to meet the non-SNMP management environments. For example, when out-of-band IP management is used via a separate management interface (e.g., for a device that does not support in-band IP access), a uniform way to indicate how to contact the device for management is needed. RFC4088 (2005) defines a URI (Uniform Resource Identifiers) scheme so that SNMP can be designated as the protocol used for management. The scheme also allows a URI to designate one or more MIB object instances.

To integrate the management of IEEE 802 networks, RFC 4789 (2006)(this document obsoletes RFC 1089) specifies how SNMP messages can be transmitted directly over IEEE 802 networks.

To meet the SNMPv3 requirements that an application needs to localize the identifier (snmpEngineID) of the remote SNMP protocol engine in order to retrieve or manipulate objects maintained on the remote SNMP entity, RFC 5343 (2008) introduces a well-known localEngineID and a discovery mechanism that can be used to learn the snmpEngineID of a remote SNMP protocol engine. The proposed mechanism is independent of the features provided by SNMP security models and may also be used by other protocol interfaces providing access to managed objects.

To deal with the large traffic measurement, RFC 5345 (2008) describes an approach to carrying out large-scale SNMP traffic measurements in order to develop a better understanding of how the SNMP is used in real-world production networks. It describes the motivation, the measurement approach, and the tools and data formats needed to carry out such an application.

RFC 5590 (2009) defines a subsystem, extending the Simple Network Management Protocol (SNMP) architecture defined in RFC 3411. As work is being done to expand the transports to include secure transports, such as the Secure Shell (SSH) Protocol and Transport Layer Security Transport Subsystem for the SNMP.

RFC 5591 (2009) describes a Transport Security Model for the SNMP.

RFC 5592 (2009) describes a Transport Model for the Simple Network Management Protocol (SNMP), using the Secure Shell (SSH) protocol. It also defines a portion of the Management Information Base (MIB) for use with network management protocols in TCP/IP-based internets. In particular, it defines objects for monitoring and managing the Secure Shell Transport Model for SNMP.

RFC 5608 (2009) describes the use of a Remote Authentication Dial-In User Service (RADIUS) authentication and authorization service with SNMP secure transport models to authenticate users and authorize creation of secure transport sessions.

3.3.3   Comparison of SNMP and CMIP

The common between SNMP and CMIP is that both perform management operations. They can move management information from one system to another, so the information can be retrieved from a device modified by the manager and returned to the device. Retrieved information can also be used to detect a malfunction of the device.

However, CMIP and SNMP are complementary and serve two different purposes. The differences between CMIP and SNMP are present in a wide number of areas.

Advantages of SNMP

SNMP is commonly considered to be a quickly designed "band-aid" solution to internetwork management difficulties while other larger and better protocols were being designed. However, no better choice became available, and SNMP soon became the network management protocol of choice.

- The largest advantage of SNMP over CMIP is that its design is simple, so it is as easy to use on a small network as well as on a large one, with ease of setup, and lack of stress on system resources. Also, the simple design makes it simple for the user to program system variables that they would like to monitor.

- Another major advantage to SNMP is that is in wide use today around the world. Because of its development during a time when no other protocol of this type existed, it became very popular, and is a built in protocol supported by most major vendors of networking hardware, such as hubs, bridges, and routers, as well as major operating systems.

- An SNMP implementation is smaller and faster than a full CMIP one. In particular SNMP requires an underlying service based on an unreliable datagram transport service and, therefore, it can be used on a wide range of datagram implementations, like UDP in TCP/IP.

- Actually, an SNMP-like protocol is the only solution for devices having a small amount of resources in terms of memory, processing power, etc. Such an architecture is particularly adequate for local environment where a number of stations, PCs and other devices such as printers, modems, etc., must be managed. Consistency at the level of management is implicitly maintained by the local character of the environment.

SNMP is by no means a perfect network manager. The first problem realized by most companies is that there are some rather large security problems related with SNMP. Any decent hacker can easily access SNMP information, giving them any information about the network and the ability to potentially shut down systems on the network. The latest version of SNMP has added some security measures that were left out of SNMP, to combat the 3 largest problems plaguing SNMP: (1) Privacy of Data (to prevent intruders from gaining access to information carried along the network), (2) authentication (to prevent intruders from sending false data across the network), and (3) access control (which restricts access of particular variables to certain users, thus removing the possibility of a user accidentally crashing the network).

Advantages of CMIP

CMIP is part of the OSI application layer and requires the implementation of a full OSI connection-oriented protocol stack. CMIP was designed to be better than SNMP in every way by repairing all flaws, and expanding on what was good about it, making it a bigger and more detailed network manager.

- It is generally agreed that CMIP is more powerful than SNMP in terms of functionality, it is essential to

  - lighten and facilitate the network manager task,
  - maintain the conductivity between local sites,
  - minimize the cost for management in terms of networking resources.

- The variables of CMIP protocol is not only relay information to and from the terminal (as in SNMP), but they can also be used to perform tasks that would be impossible under SNMP. For instance, if a terminal on a network cannot reach the fileserver a pre-determined amount of times, then CMIP can notify appropriate personnel of the event. With SNMP, however, a user would have to specifically tell it to keep track of unsuccessful attempts to reach the server, and then what to do when that variable reaches a limit. CMIP therefore results in a more efficient management system, and less work is required from the user to keep updated on the status of the network.

- CMIP also contains the security measures left out by SNMP. Because of the large development budget, when it becomes available, CMIP will be widely used by the government, and the corporations that funded it.

- CMIP tends to share equally the resources required for the network management purposes between the manager station and the agent. CMIP includes a number of functions, most of them being optional and negotiable. CMIP gives us the building boxes to design a powerful network management architecture, which is particularly appropriate for the management of networks, including a large number of switches.

However, CMIP requires about five to ten times the system resources that are needed for SNMP. In other words, very few systems in the world would be able to handle a full implementation on CMIP without undergoing massive network modifications. This disadvantage has no inexpensive fix to it. The other flaw in CMIP is that it is very difficult to program. Its complex nature requires so many different variables that only a few skilled programmers are able to use it to its full potential.

Technical Comparison of CMIP and SNMP

Considering the above information, one can see that both management systems have their advantages and disadvantages. A detailed comparison of SNMP and CMIP is denoted in Table 3.2 [KV97].

3.3.4 Internet Protocol Flow Information Export (IPFIX)

Internet Protocol Flow Information Export (IPFIX) is for a common, universal standard of export for Internet Protocol flow information from routers, probes, and other devices that is used by mediation systems, accounting/billing systems, and network management systems to facilitate services such as measurement, accounting, and billing. The IPFIX standard will define how IP flow information is to be formatted and transferred from an exporter to a collector. Previously, many data network operators were relying on the proprietary Cisco Systems Netflow standard for traffic-flow information export. The IPFIX standards requirements were outlined in the original RFC 3917. The working group chose Cisco Netflow Version 9 as the basis for IPFIX. The working group submitted the IPFIX Protocol Specification to the IESG for approval in 2006.

Recently advances in IPFIX can be found in RFC 3955, 5101, 5103, 5153, 5471, 5472, 5473, and 5610.

The following figure shows a typical architecture of information flow in an IPFIX architecture (see Figure 3.18):

A Metering Process collects data packets at an Observation Point, optionally filters them, and aggregates information about these packets. Using the IPFIX protocol, an Exporter then sends this information to a Collector. Exporters and Collectors are in a many-to-many relationship: One Exporter can send data to many Collectors and one Collector can receive data from many Exporters.

IPFIX considers a flow to be any number of packets observed in a specific timeslot and sharing a number of properties, e.g., "same source, same destination, same protocol." Using IPFIX, devices like routers can inform a central monitoring station about their view of a potentially larger network.

Table 3.2: A Detailed Comparison of SNMP and CMIP

Feature	SNMP	CMIP
Agent intelligence	Simple agents	Complex and powerful agents
Environment	TCP/IP	OSI
Manager intelligence	Polling-based	Agents can filter events, accept "action" commands (event-based)
Bandwidth requirements	Excessive polling can result in high bandwidth requirements	Larger messages can result in high bandwidth consumption
Code size	Small	Fairly large
Cost per network element	Inexpensive to implement since SNMP, UDP and IP have no or low license fees	License fees are of order few US$1000
Scalability	Polling methods require careful tuning of network	Event driven creates a more scalable solution for large reliable networks
Security	Secure SNMPv2 offers authentication and access control, including alarm	CMIP offers security as well as audit trial service
Naming	Local naming requires more information to make it globally unique	Uses X.500 naming to facilitate manager to manager distribution
Communications	Connectionless datagram using UDP/IP; places burden on manager for recovery	Connection oriented upper layers using OSI sever layer stack
Data retrieval	Simple reads and writes	Structured queries
Perspective on data	Flat file	Object oriented
Modes of operation	Unconfirmed	Confirmed
Redundancy of management	Agents may send traps to multiple managers	Management services can be distributed to multiple destinations
Representation	ASN.1	Object oriented
Data synchronization	One row at a time	Scoped GET provides agent synchronization of data

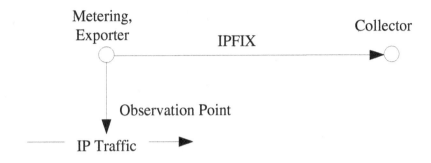

Figure 3.18: IPFIX Architecture

IPFIX is a push protocol, i.e., each sender will periodically send IPFIX messages to configured receivers without any interaction by the receiver.

The actual makeup of data in IPFIX messages is to a great extent up to the sender. IPFIX introduces the makeup of these messages to the receiver with the help of special Templates. The sender is also free to use user-defined data types in its messages, so the protocol is freely extensible and can adapt to different scenarios.

IPFIX prefers the Stream Control Transmission Protocol as its transport layer protocol, but also allows the use of the Transmission Control Protocol or User Datagram Protocol.

A simple information set sent via IPFIX might look like this (Table 3.3):

Table 3.3: IPFIX Information

Source	Destination	Packets
192.168.0.201	192.168.0.1	235
192.168.0.202	192.168.0.1	42

This information set would be sent in the following IPFIX message (Table 3.4):

As can be seen, the message contains the IPFIX header and two IPFIX Sets: One Template Set that introduces the build-up of the Data Set used, as well as one Data Set, which contains the actual data. Because the Template Set is buffered in Collectors it will not need to be transmitted in subsequent messages.

An IPFIX Device consists of a set of co-operating processes that implement the functional blocks described in the previous section. Alternatively, an IPFIX Device can be viewed simply as a network entity, which implements the IPFIX protocol. At the IPFIX Device, the protocol functionality resides

Table 3.4: IPFIX Message

Bits 0...15	Bits 16...31
Version=0x000a	Message Length =64 Byte
Export Timestamp = 2008-11-23 23:59:34	
Sequence Number = 0	
Source ID = 12345678	
Set ID =2 (Template)	Set Length = 20 Byte
Template ID =256	Number of Field =3
Typ =sourceIPv4Address	Field Length = 4 Byte
Typ =destinationIPv4Address	Field Length = 4 Byte
Typ =packetDeltaCount	Field Length = 8 Byte
Set ID = 256 (Data Set using Template 256)	Set Length = 24 Byte
Record 1, Field 1 = 192.168.0.201	
Record 1, Field 2 = 192.168.0.1	
Record 1, Field 3 = 235 Packets	
Record 2, Field 1 = 192.168.0.202	
Record 2, Field 2 = 192.168.0.1	
Record 2, Field 3 = 42 Packets	

in the Exporting Process. The IPFIX Exporting Process gets Flow Records from a Metering Process, and sends them to the Collector(s).

At a high level, an IPFIX Device performs the following tasks:

1. Encode Control Information into Templates.

2. Encode packets observed at the Observation Points into Flow Records.

3. Packetize the selected Templates and Flow Records into IPFIX Messages.

4. Send IPFIX Messages to the Collector.

The IPFIX protocol communicates information from an IPFIX Exporter to an IPFIX Collector. That information includes not only Flow Records, but also information about the Metering Process. Such information (referred to as Control Information) includes details of the data fields in Flow Records. It may also include statistics from the Metering Process, such as the number of packets lost (i.e., not metered).

For security, IPFIX should include more efficient strategies to improve the protection and security of traffic flows.

The data modeling part of the IPFIX framework may be involved. The informal ad-hoc notation makes it difficult to build tools and it would be desirable to achieve some level of integration with the data models used by other management protocols [Sch07].

3.3.5 Network Configuration Protocol (NETCONF)

NETCONF is a network management protocol developed in the IETF by the Netconf working group. It was published as RFC 4741.

The NETCONF protocol provides mechanisms to install, manipulate, and delete the configuration of network devices. It also can perform some monitoring functions. It uses an Extensible Markup Language (XML) based data encoding for the configuration data as well as the protocol messages. The NETCONF protocol operations are realized on top of a simple Remote Procedure Call (RPC) layer. This in turn is realized on top of the transport protocol.

The NETCONF protocol uses a remote procedure call (RPC) paradigm. A client encodes an RPC in XML and sends it to a server using a secure, connection-oriented session. The server responds with a reply encoded in XML. The contents of both the request and the response are fully described in XML (the Extensible Markup Language) DTDs (Document Type Definitions) or XML schemas, or both, allowing both parties to recognize the syntax constraints imposed on the exchange.

A key aspect of NETCONF is that it allows the functionality of the management protocol to closely mirror the native functionality of the device. This reduces implementation costs and allows timely access to new features. In addition, applications can access both the syntactic and semantic content of the device's native user interface.

NETCONF allows a client to discover the set of protocol extensions supported by a server. These "capabilities" permit the client to adjust its behavior to take advantage of the features exposed by the device. The capability definitions can be easily extended in a noncentralized manner. Standard and non-standard capabilities can be defined with semantic and syntactic rigor.

The NETCONF protocol is a building block in a system of automated configuration. XML is the lingua franca of interchange, providing a flexible but fully specified encoding mechanism for hierarchical content. NETCONF can be used in concert with XML-based transformation technologies, such as XSLT (Extensible Stylesheet Language Transformations), to provide a system for automated generation of full and partial configurations. The system can query one or more databases for data about networking topologies, links, policies, customers, and services. This data can be transformed using one or more XSLT scripts from a task-oriented, vendor-independent data schema into a form that is specific to the vendor, product, operating system, and software release. The resulting data can be passed to the device using the NETCONF protocol.

NETCONF can be conceptually partitioned into four layers:

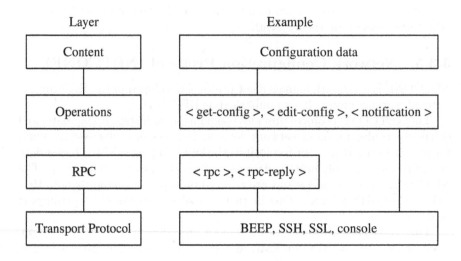

Figure 3.19: Layers of NETCONF

1. The transport protocol layer provides a communication path between the client and server. NETCONF can be layered over any transport protocol that provides a set of basic requirements.

2. The RPC layer provides a simple, transport-independent framing mechanism for encoding RPCs.

3. The operations layer defines a set of base operations invoked as RPC methods with XML-encoded parameters.

4. The content layer is outside the scope of this document. Given the current proprietary nature of the configuration data being manipulated, the specification of this content depends on the NETCONF implementation. It is expected that a separate effort to specify a standard data definition language and standard content will be undertaken.

Recently, RFC 5381 (2008) describes the development of a SOAP (Simple Object Access Protocol)-based NETCONF (Network Configuration Protocol) client and server. It describes an alternative SOAP binding for NETCONF that does not interoperate with an RFC 4743 conformant implementation making use of cookies on top of the persistent transport connections of HTTP. When SOAP is used as a transport protocol for NETCONF, various kinds of development tools are available. By making full use of these tools, developers can significantly reduce their workload. The authors developed an NMS (Network Management System) and network equipment that can deal with NETCONF messages sent over SOAP.

RFC 5539 (2009) describes how to use the Transport Layer Security (TLS) protocol to secure NETCONF exchanges.

The downside of NETCONF, as it is defined today, are the missing pieces, namely, a lack of standards for data modeling and the lack of a standardized access control model. Since NETCONF implementations are well underway, it can be expected that these shortcomings will be addressed in the next few years and NETCONF might become a powerful tool for network configuration and trouble shooting [Sch07].

3.3.6 Syslog

The term "syslog" is often used for both the actual syslog protocol, as well as the application or library sending syslog messages.

The syslog protocol provides a transport to allow a machine to send event notification messages across IP networks to event message collectors – also known as syslog servers. Since each process, application, and operating system was written somewhat independently, there is little uniformity to the content of syslog messages. For this reason, no assumption is made upon the formatting or contents of the messages. The protocol is simply designed to transport these event messages. In all cases, there is one device that originates the message. The syslog process on that machine may send the message to a collector. No acknowledgement of the receipt is made.

One of the fundamental tenets of the syslog protocol and process is its simplicity. No stringent coordination is required between the transmitters and the receivers. Indeed, the transmission of syslog messages may be started on a device without a receiver being configured, or even actually physically present. Conversely, many devices will most likely be able to receive messages without explicit configuration or definitions. This simplicity has greatly aided the acceptance and deployment of syslog.

Syslog is a client/server protocol: the syslog sender sends a small (less than 1KB) textual message to the syslog receiver. The receiver is commonly called "syslogd," "syslog daemon" or "syslog server." Syslog messages can be sent via UDP and/or TCP. The data are sent in cleartext; although not part of the syslog protocol itself, an SSL wrapper can be used to provide for a layer of encryption through SSL/TLS.

Syslog is typically used for computer system management and security auditing. While it has a number of shortcomings, syslog is supported by a wide variety of devices and receivers across multiple platforms. Because of this, syslog can be used to integrate log data from many different types of systems into a central repository.

Syslog has since become the standard logging solution on Unix and Linux systems; there have also been a variety of syslog implementations on other operating systems and is commonly found in network devices such as routers.

Syslog has proven to be an effective format to consolidate logs with, as there are many open source and commercial tools for reporting and analysis.

An emerging area of managed security services is the collection and analysis of syslog records for organizations. The MSSPs are able to apply artificial intelligence algorithms to detect patterns and alert customers of problems.

The syslog protocol presents a spectrum of service options for provisioning an event-based logging service over a network. Each option has associated benefits and costs. Accordingly, the choice as to what combination of options is provisioned is both an engineering and administrative decision.

The syslog service supports three roles of operation: device, relay, and collector.

Devices and collectors act as sources and sinks, respectively, of syslog entries.

The relationship between devices and collectors is potentially many- to-many. I.e., a device might communicate with many collectors; similarly, a collector might communicate with many devices.

A relay operates in both modes, accepting syslog entries from devices and other relays and forwarding those entries to collectors and other relays. See Figure 3.20.

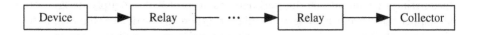

Figure 3.20: The Relationship of Device, Relay, and Collector for Syslog

As shown, more than one relay may be present between any particular device and collector.

A relay may be necessary for administrative reasons. For example, a relay might run as an application proxy on a firewall. Also, there might be one relay per company department, which authenticates all the devices in the department, and which in turn authenticates itself to a company-wide collector.

A relay can also serve to filter messages. For example, one relay may collect the syslog information from an entire web server farm, summarizing hit counts for report generation, forwarding "page not found" messages (indicating a possible broken link) to a collector that presents it to the webmaster, and sending more urgent messages (such as hardware failure reports) to a collector that gateways them to a pager. A relay may also be used to convert formats from a device's output to a collector's input.

It should be noted that a role of device, relay, or collector is relevant only to a particular BEEP channel. A single server can serve as a device, a relay, and a collector, all at once, if so configured. It can even serve as a relay and a collector to the same device at the same time using different BEEP channels over the same connection-oriented session; this might be useful to collect status yet relay urgent error messages.

To provide reliable delivery when realizing the syslog protocol, two BEEP profiles are defined. BEEP is a generic application protocol framework for connection-oriented, asynchronous interactions. Within BEEP, features such as authentication, privacy, and reliability through retransmission are provided.

- The RAW profile is designed to provide a high-performance, low- impact footprint, using essentially the same format as the existing UDP-based syslog service.

- The COOKED profile is designed to provide a structured entry format, in which individual entries are acknowledged (either positively or negatively).

Note that both profiles run over BEEP. BEEP defines "transport mappings," specifying how BEEP messages are carried over the underlying transport technologies. Both the RAW and COOKED profile provide reliable delivery of their messages. The choice of profile is independent of the operational roles discussed above.

The syslog protocol has been very successful in achieving wide spread deployment.

RFC 3195, 5424, 5425, 5426, 5427, 5675, and 5676 present more details about syslog protocols and recent advances.

3.3.7   Other Protocols Related to Network Management

Ping

Ping is commonly used to check connectivity between devices. It is the most common protocol used for availability polling. It can also be used for troubleshooting more complex problems in the network. Ping uses the Internet Control Message Protocol (ICMP) Echo and Echo Reply packets to determine whether one IP device can talk to another. Most implementations of ping allow users to vary the size of the packet.

Traceroute

Traceroute is most commonly used to troubleshoot connectivity issues. If we know that we cannot reach a host from another host, traceroute will show whether the connectivity loss exists at one of the intermediate routers. Traceroute determines that the destination has been reached when it receives an ICMP destination port unreachable message. Note that we are actually discovering the path that the ICMP timeout messages are taking when they come back.

Terminal Emulators

Terminal emulators are used for many purposes in network management, including users' accesses to network devices. Obviously, access is useful for configuring and troubleshooting devices. There are also times when information or operations on network devices is not available through SNMP and scripts must be written to access this information or capability through terminal access. Telnet is the traditional way of obtaining terminal emulation access to network devices.

3.4 Evolution in Network Management Functions

3.4.1 FCAPS Network Management Functions

FCAPS is the ISO Telecommunications Management Network model and framework for network management. FCAPS is an acronym for Fault, Configuration, Accounting, Performance, Security, which are the management categories into which the ISO model defines network management tasks.

Fault Management

Fault Management refers to the set of functions that detect, isolate, and correct malfunctions in a telecommunications network, compensate for environmental changes, and include maintaining and examining error logs, accepting and acting on error detection notifications, tracing and identifying faults, carrying out sequences of diagnostics tests, correcting faults, reporting error conditions, and localizing and tracing faults by examining and manipulating database information. When a fault or event occurs, a network component will often send a notification to the network operator using a protocol such as SNMP. An alarm is a persistent indication of a fault that clears only when the triggering condition has been resolved. A current list of problems occurring on the network component is often kept in the form of an active alarm list such as is defined in RFC 3877, the Alarm MIB. A list of cleared faults is also maintained by most network management systems.

A fault management console allows a network administrator or system operator to monitor events from multiple systems and perform actions based on this information. Ideally, a fault management system should be able to correctly identify events and automatically take action, either launching a program or script to take corrective action, or activating notification software that allows a human to take proper intervention (i.e., send e-mail or SMS text to a mobile phone). Some notification systems also have escalation rules that will notify a chain of individuals based on availability and severity of alarm.

There are two primary ways to perform fault management - these are active and passive. Passive fault management is done by collecting alarms from devices (normally via SNMP) when something happens in the devices. In this mode, the fault management system only knows if a device it is monitoring is intelligent enough to generate an error and report it to the management tool. However, if the device being monitored fails completely or locks up, it won't throw an alarm and the problem will not be detected. Active fault management addresses this issue by actively monitoring devices via tools such as PING to determine if the device is active and responding. If the device stops responding, active monitoring will throw an alarm showing the device as unavailable and allows for the proactive correction of the problem.

Fault management can typically be broken down into three basic steps, namely [ANS94]:

1. **Fault Detection** [BCF94]: the process of capturing on-line indications of disorder in networks provided by malfunctioning devices in the form of alarms. Fault detection determines the root cause of a failure. In addition to the initial failure information, it may use failure information from other entities in order to correlate and localize the fault.

2. **Fault Isolation** [BCF+95] [KG97] (also referred to as fault localization, event correlation, and root cause analysis): a set of observed fault indications is analyzed to find an explanation of the alarms. This stage includes identifying the cause that lead to the detected failure in case of fault-propagation and the determination of the root cause.

   The fault localization process is of significant importance because the speed and accuracy of the fault management process are heavily dependent on it.

3. **Fault Correction**: is responsible for the repair of a fault and for the control of procedures that use redundant resources to replace equipment or facilities that have failed.

Ideally, fault management will include all three steps starting with the detection, but most often only the fault-detection is implemented because of the complexity in providing general fault isolation and fault correction procedures. As mentioned above, fault management is a complex process and consists of two diagnosis steps (fault detection and fault isolation) and one planning step (fault correction).

Configuration Management

Configuration Management refers to the process of initially configuring a network and then adjusting it in response to changing network requirements. This function is perhaps the most important area of network management because improper configuration may cause the network to work suboptimally or to not work at all. An example is the configuration of various parameters on a network interface. The basic tasks of configuration management include:

- Facilitate the creation of controls

- Monitor and enforce baseline standards for specific hardware and software

- Store the configuration data and maintain an up-to-date inventory of all network components

- Log and report changes to configurations, including user identity

Network configuration management have to deal with detailed operation on network hardware configuration, network software configuration, and maintenance management.

1. Network hardware configuration management

   Hardware configuration management is the process of creating and maintaining an up-to-date record of all the components of the infrastructure, including related documentation. Its purpose is to show what makes up the infrastructure and illustrate the physical locations and links between each item, which are known as configuration items.

   The scope of hardware configuration management is assumed to include:

   - physical client and server hardware products and versions
   - operating system software products and versions
   - application development software products and versions
   - technical architecture product sets and versions as they are defined and introduced
   - live documentation
   - networking products and versions
   - live application products and versions
   - definitions of packages of software releases
   - definitions of hardware base configurations
   - configuration item standards and definitions

2. Software configuration management

   Network software configuration management (SCM) process includes:

   - Configuration identification
     Configuration identification is the process of identifying the attributes that define every aspect of a configuration item. These attributes are recorded in configuration documentation and baselined. Baselining an attribute forces formal configuration change control processes to be effected in the event that these attributes are changed.

- Configuration change control

  Configuration change control is a set of processes and approval stages required to change a configuration item's attributes and to re-baseline them.

- Configuration status accounting

  Configuration status accounting is the ability to record and report on the configuration baselines associated with each configuration item at any moment of time.

- Configuration authentication

  Configuration audits are broken into functional and physical configuration audits. They occur either at delivery or at the moment of effecting the change. A functional configuration audit ensures that functional and performance attributes of a configuration item are achieved, while a physical configuration audit ensures that a configuration item is installed in accordance with the requirements of its detailed design documentation.

Accounting Management

Accounting Management refers to gather usage statistics for users. Using the statistics the users can be billed and usage quota can be enforced. The measurement of network utilization parameters enables individual or group uses on the network to be regulated appropriately. Such regulation minimizes network problems (because network resources can be apportioned based on resource capacities) and maximizes the fairness of network access across all users. Listed below are the primary tasks/ services performed by accounting management accountants. The degree of complexity relative to these activities are dependent on the experience level and abilities of any one individual:

- Network resource utilization

- Variance analysis

- Rate and volume analysis

- Business metrics development

- Price modeling

- Product profitability

- Cost analysis

- Cost benefit analysis

- Cost-volume-profit analysis

- Client profitability analysis

- Billing information

- Strategic planning

- Strategic management advise

- Sales and financial forecasting

- Resource allocation and utilization

- Regulate users or groups

- Help keep network performance at an acceptable level

RADIUS, TACACS, and Diameter are examples of protocols commonly used for accounting.

Performance Management

Performance Management refers to monitoring network utilization, end-to-end response time, and other performance measures at various points in a network. The results of the monitoring can be used to improve the performance of the network.

In network performance management, 1) a set of functions that evaluate and report the behavior of telecommunications equipment and the effectiveness of the network or network element and 2) a set of various subfunctions, such as gathering statistical information, maintaining and examining historical logs, determining system performance under natural and artificial conditions, and altering system modes of operation.

Examples of performance variables that might be provided include network throughput, user response times, and line utilization.

There are two central tasks involved in performance management: monitoring (performance analysis) and control (capacity planning).

Monitoring consists in obtaining the utilization and error rates of current network links and network devices, collecting information about traffic (error rates, throughput, loss, utilization, collisions, volume, matrix), services (protocols, overhead, matrix, response time), and resources (CPU utilization, memory available, disk utilization, ports) and comparing this information with the normal and/or desirable values for each.

Controlling consists in making baseline the utilization metrics and isolate any existing performance problems, taking actions to plan or modify network configurations and capacities to meet the desirable performance.

Security Management

Security Management refers to the set of functions that protects networks and systems from unauthorized access by persons, acts, or influences and that includes many subfunctions, such as creating, deleting, and controlling security services and mechanisms; distributing security-relevant information; reporting security-relevant events; controlling the distribution of cryptographic keying material; and authorizing subscriber access, rights, and privileges.

A security management subsystem, for example, can monitor users logging on to a network resource and can refuse access to those who enter inappropriate access. The basic tasks of security management include:

- Identify sensitive information or devices and access control to network resources by physical and logical means

- Implement a network intrusion detection scheme to enhance perimeter security

- Protect the sensitive information by configuring encryption policies

- Possibly respond automatically to security breaches and attempts with pre-defined operations

Management tools such as information classification, risk assessment, and risk analysis are used to identify threats, classify assets, and to rate system vulnerabilities so that effective control can be implemented.

The FCAPS model is useful for understanding the goals and requirements of Network Management, and also helps to build a foundation for understanding the significance of Network Management to compliance efforts.

3.4.2 Expanded Network Management Functions

SLA (Service Level Agreement) Management

1. Introduction of SLA

   A service-level agreement (SLA) is a negotiated agreement between two parties where one is the customer and the other is the service provider. This can be a legally binding formal or informal "contract" (see internal department relationships).

   The SLA records a common understanding about services, priorities, responsibilities, guarantees, and warranties. Each area of service scope should have the "level of service" defined. The SLA may specify the levels of availability, serviceability, performance, operation, or other attributes of the service such as billing. The "level of service" can also be specified as "target" and "minimum," which allows customers to informed what to expect (the minimum), whilst providing a measurable (average) target value that shows the level of organization performance. In some contracts, penalties may be agreed in the case of noncompliance of the SLA (but see "internal" customers below). It is important to note that the "agreement" relates to the services the customer receives, and not how the service provider delivers that service.

SLAs have been used since late 1980s by fixed line telecom operators as part of their contracts with their corporate customers. This practice has spread such that now it is common for a customer to engage a service provider by including a service-level agreement in a wide range of service contracts, in practically all industries and markets. Internal departments in larger organizations (such as IT, HR, and Real Estate) have adopted the idea of using service-level agreements with their "internal" customers – users in other departments within the same organization. One benefit of this can be to enable the quality of service to be benchmarked with that agreed across multiple locations or between different business units. This internal benchmarking can also be used to market test and provide a value comparison between an in-house department and an external service provider.

Service-level agreements are by their nature "output" based – the result of the service as received by the customer is the subject of the "agreement". The (expert) service provider can demonstrate their value by organizing themselves with ingenuity, capability, and knowledge to deliver the service required, perhaps in an innovative way. Organizations can also specify the way the service is to be delivered, through a specification (a service-level specification) and using subordinate "objectives" other than those related to the level of service. This type of agreement is known as an "input" SLA. This latter type of requirement has become obsolete as organizations become more demanding and shift the delivery methodology risk on to the service provider.

2. SLA Management

SLAs are based on fixed service and performance agreements between customers and suppliers and create transparency for both parties in terms of performance and costs. Specific SLAs are used to define the type, scope, and quality of services and to check that specifications are met. As SLAs also include potential sanctions for the event that agreed service parameters are not met, the specifications made in them have a significant effect on the commercial success of a company providing services for a customer. In the context of service-oriented architectures, the benefits of successful service-level management can be described as follows [SDR04]:

- The number of conflict situations within supplier relationships can be reduced, resulting in enhanced customer satisfaction.
- The resources used in order to render the service (hardware, personnel, licenses) can be distributed at a detailed level by the provider and therefore used in such a way as to optimize costs.
- Problems can be identified speedily by service-level monitoring and the associated cause determined.

- Costs can be made more transparent; on the one hand, the customer only wants to pay for services actually used while, on the other hand, plausible pricing can be guaranteed.

[KM08] provides a SLA management process model, see Figure 3.21.

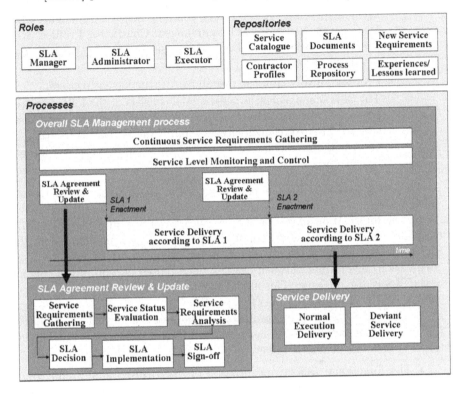

Figure 3.21: SLA Management Process Model

The model consists of:

- Roles involved in the SLA management process.
  The roles involved in the SLA management process include Service Level Manager, Service Level Administrator, and Service Level Executor.
  SLM Manager owns the SLA management process. He defines, establishes and manages it within the organization.
  SLA Administrator is responsible for scheduling and arranging the SLA reviews, documenting changes to SLAs, and for supporting the SLA Manager with various administrative tasks.
  SLA Executor provides daily service to service consumers according to the service requirements as agreed upon in the SLAs.

- Repositories storing information and knowledge required for managing the SLAs.

  Information required for monitoring service and managing collaboration needs be recorded in an SLA repository. An effective repository is a primary key success factor for an effective SLA management process. A repository contains Service Catalogue, SLA Documents, New Service Requirements, Contractor Profiles, SLA Process Repository, and Experiences and Lessons Learned. Below, we briefly describe them.

- Processes involved in the overall SLA management process.

  The SLA management process model consists of the following processes: (1) Service Delivery, (2) Service Level Monitoring and Control, (3) SLA Agreement, Review and Update, and (4) Continuous Service Requirements Gathering. All these processes are performed concurrently.

The open issues for SLA management focus on these particular points [MMB+02]:

1) The information model of SLS (Service Level Specification). The SLS template or information model is not clearly defined. It is mandatory for the SLS to be defined by a template or an information model in order to allow cooperation/negotiation between entities. The representation of the content of the SLS is also an issue. This model should also cope with various types of service (3GPP, Multimedia, IP-VPN, etc.).

2) The SLS negotiation protocol. The negotiation allows cooperation/negotiation between OSS, i.e., a Service Provider (Operator) and a customer (which can be another operator) and can be divided into three aspects:

- The Functional aspect: negotiation of SLSs (provisioning and assurance aspect), modification of an implemented SLS, information about an implemented SLS (state, performance, etc);

- The Security aspect: authentication, access control, integrity and confidentiality;

- The Inter-domain aspect: SLS negotiation between distinct administrative domains. The scope of SLSs is limited to a domain or may cover several administrative domains.

  3) The end-to-end point of view. This induces that agreements have to be established between providers, and between providers and customers. Therefore, it is necessary to establish Out-Sourcing agreements (an SLA) with other networks providers to lease part of their network, that can be for instance defined as leased line, VPN, etc.

The automatic Management of the SLA/QoS requires the mapping of the SLA requirement into technical configuration of network equipment and the specification of tools to generate QoS parameters from SLA. The SLA monitoring must be improved in order to determine service performance measurements relevant for efficient SLA monitoring, to manage the network to maintain the SLA requirements (equipment reconfiguration), and to optimize the network performance and the network usage.

Change Management

Change Management is an IT Service Management discipline. The objective of Change Management in this context is to ensure that standardized methods and procedures are used for efficient and prompt handling of all changes to controlled IT infrastructure, in order to minimize the number and impact of any related incidents upon service. Changes in the IT infrastructure may arise reactively in response to problems or externally imposed requirements, e.g., legislative changes, or proactively from seeking imposed efficiency and effectiveness or to enable or reflect business initiatives, or from programs, projects, or service improvement initiatives. Change Management can ensure standardized methods, processes, and procedures are used for all changes, facilitate efficient and prompt handling of all changes, and maintain the proper balance between the need for change and the potential detrimental impact of changes.

Change Management would typically comprise the raising and recording of changes, assessing the impact, cost, benefit, and risk of proposed changes, developing business justification and obtaining approval, managing and co-ordinating change implementation, monitoring and reporting on implementation, reviewing and closing change requests.

ITIL (The Information Technology Infrastructure Library) defines the change management process this way:

The goal of the Change Management process is to ensure that standardized methods and procedures are used for efficient and prompt handling of all changes, in order to minimize the impact of change-related incidents upon service quality, and consequently improve the day-to-day operations of the organization.

Change management is responsible for managing change process involving:

- Hardware

- Communications equipment and software

- System software

- All documentation and procedures associated with the running, support, and maintenance of live systems.

Any proposed change must be approved in the change management process. While change management makes the process happen, the decision authority is the Change Advisory Board (CAB), which is made up for the most part of people from other functions within the organization. The main activities of the change management are:

- Filtering changes

- Managing changes and the change process

- Chairing the CAB and the CAB/Emergency committee

- Reviewing and closing of Requests for Change (RFCs)

- Management reporting and providing management information

Figure 3.22 depicts the process of change management.

Situation Management

Situation management (SM) is a goal-directed close-loop process of sensing, reasoning, perceiving, prediction, and affecting of event-driven dynamic systems, where evolving situations determine the overall behavior of the system. The overall motivation for SM is achieving using the system a desired goal situation within the predefined limits. Situation management can improve the network management functions in modern dynamic networks and integrated IT systems.

Situation management as a management concept and technology is taking roots from disciplines, including situation awareness, situation calculus, and situation control. Critical aspects of situation management include managing and controlling sources of information, processing real-time or near real-time streams of events, representing and integrating low-level events and higher-level concepts, multi-source information fusion, information presentations that maximize human comprehension, and reasoning about what is happening and what is important. Furthermore, commanders will require management support systems that include control over their current command options, prediction of probable situation evolutions, and analysis of potential threats and vulnerabilities [JLM+05].

As a rule, situations often involve a large number of dynamic objects that change states in time and space and engage each other into fairly complex spatio-temporal relations. From a management viewpoint it is important to understand the situations in which these objects participate, to recognize emerging trends and potential threats, and to undertake required actions. Understanding dynamic situations requires complex cognitive modeling, the building of ontologies, and continuous collection, filtering, and fusion of sensor, intelligence, database, Internet-based, and related information sources.

The primary components of situation management are sensing, reasoning, and controlling. See Figure 3.23.

SM model includes several levels of a sense/reason/control loop. Typically, the data at the lower levels are close to being "raw," i.e., what is immediately

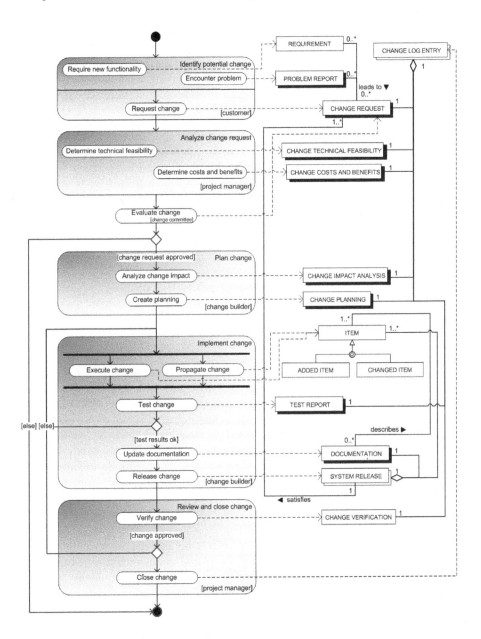

Figure 3.22: Change Management Process

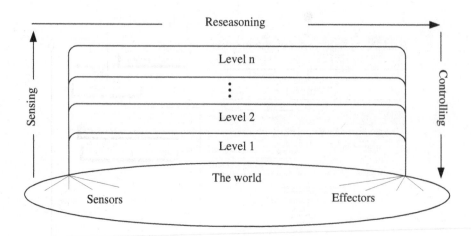

Figure 3.23: Model of Situation Management

returned by a sensor. Such data might be examined immediately and issue control instructions for effectors, or else simply passed to an operator for further processing. Note that such raw data could include images, text, electronic signals, etc.

The high-level of SM has to do with reasoning, knowledge, and understanding. Consequently, several important questions are generated: What is a situation? How do we represent a situation? What inputs are needed to identity a situation? How do we represent the degree of certainty or confidence in an identified situation?

Some researchers present a detailed situation management framework [JBL07]. See Figure 3.24.

There are several important driving forces behind the advancement of the filed of SM, including the need for an integrated management of complex dynamic systems, progress in several important enabling technologies, and the emergence of new critical applications such as operational battlefield management, disaster situation management, homeland security applications, and the management of complex dynamic systems and networks. At the same time SM faces several challenges, such as integration of computational and symbolic

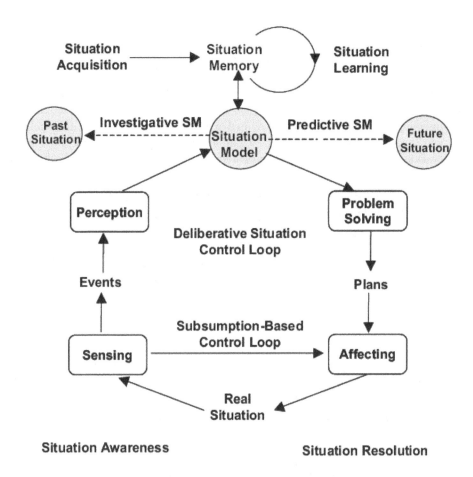

Figure 3.24: Situation Management Framework

reasoning processes, development of situation modeling languages, increasing the effectiveness of SM models, tools, and platforms, and the development of effective methods of situational learning.

From the viewpoint of application, situation management does not replace existing technologies. Situation management enhances and optimizes existing systems and technologies with added functions without disrupting the workflow.

As an example, a situation management system integrates discrete IT systems (such as closed-circuit television, sensors, access control, biometrics) into a common platform with a unified, real-time, text, or GIS-based display. The systems that enable the quickest integration offer generic gateways for each technology and custom interfaces to provide a smooth interface for any system to communicate with any other system. An integrated system can also be scheduled and synchronized through a unified control application. In this way, the time of day, day of week, alert level, or any other set of conditions can control the behavior of multiple IT systems.

Recent innovations in situation management will help response planners and command and control teams get the most accurate picture of what is happening and what to expect. Armed with this knowledge, IT managers (such as security personnel) can best manage resources, ensure continuity of operations and, of course, optimize incident response speed and accuracy.

3.4.3 Management Application vs. Management Functionality

Network management application is to execute detailed operation on management to achieve the network management functionality. Detailed application issues and related network management functions are listed as follows:

- Acquire and maintain application Software

  - Track software versions on all relevant systems
  - Track user licenses
  - Track warrantees

- Acquire and maintain technology Infrastructure

  - Monitor and report on status of all infrastructure elements

- Enable operations

  - Ensure required network resources are available to qualified users when required
  - Monitor and report on performance levels
  - Report and record network performance that does not align with pre-defined baselines

- Install and accredit solutions and changes; manage changes

  - Implement (sometimes automatically) updates, changes, patches, fixes, etc.
  - Test the effect of changes on the network before implementation; verify that changes comply to pre-defined configurations and policies
  - Track, report, and document all changes to the network; show who made what changes, where and when
  - Authorize maintenance access

- Define and manage service levels

  - Track incidents and response times
  - Provide automated service reports

- Manage third-party services

  - Allow access control and audit functions for third-party staff

- Ensure systems security

  - Enforce authentication and access policies that protect data and systems against unauthorized access and/or change via role/user profile administration, passwords, file permissions, file encryption, properly configured firewalls, etc.
  - Detect and respond to security breaches
  - Protect data when it's transferred across the network (IP Security, wireless security)
  - Protect against viruses, Trojans, worms, spyware, and other malicious software (hardware and software solutions)
  - Track who accesses regulated data and keep logs that show when, how, and by whom such data were accessed, often at the transaction level
  - Use protective mechanisms such as Secure Sockets Layer (SSL), VPN, or other secure connection technology
  - Encrypt e-mail to protect the confidentiality of the information and ensure the authenticity and integrity of the transmission

- Manage the configuration

  - Maintain (sometimes automatically) an accurate inventory of all configurable components installed on the network
  - Audit and record configuration settings on all devices
  - Monitor and enforce baseline configuration definitions

- Manage problems and incidents

  - Provide audit activity reports for forensic analysis
  - Allow emergency access management

- Manage data

  - Retain and archive e-mail (often automatically) as per retention policies
  - Perform automated backups, data storage, and archiving for disaster recovery requirements

- Store and restore configuration data for network elements

- Manage the physical environment and operations

  - Produce activity logs
  - Track and audit all devices connected to the network
  - Produce and maintain up-to-date inventory of all devices connected to the network
  - Optimize use of resources by balancing loads

3.5 Challenges in Network Management

Future networks will probably have novel characteristics respect to today's networks. These are some examples:

- They would use flexibly and efficiently radio access, allowing ubiquitous access to broadband nomadic and mobile services;

- They will manage in real time new forms of ad-hoc communications with intermittent connectivity requirements and time-varying network topology;

- They will integrate sub-networks at the edge, such as personal and sensor networks, toward the Internet of Things for the benefit of humans;

- They will eliminate the barriers to broadband access and will enable intelligent distribution of services across multiple access technologies with centralized or distributed control;

- They will enable seamless end-to-end network and service composition and operation across multiple operators and business domains;

- Finally, to support high-quality media services and support critical infrastructure (e.g., for energy and transport), the existing Internet will be significantly enhanced or even gradually replaced.

To overcome current limitations and to address emerging trends, such as mobility, diffusion of heterogeneous nodes and devices, mass digitization, new forms of user-centered content provisioning, emergence of software as a service, and of new models of service and interaction with improved security and privacy features, networks meet increasing challenges in the services, technologies, security, and management. Consequently, the network management should meet the following challenges:

- The increasing pervasiveness of Mobility and Wireless technologies;

- The soaring number of connected devices, eventually leading to sensor networks;

- The accelerated race for processing power and memory increase, continuing to support the trend of more and more intelligence at the network periphery;

- The expected heavy increase in digitized media, user-generated content, and associated critical requirement for data search, handling, and organization.

- Location determination, as an important enabler for new categories of context aware services;

- End user provided infrastructure and services, possibly driving a user-generated infrastructure, similar to the trend towards user-generated content;

- Security and resilience of the infrastructures, associated with growing concerns for privacy in an environment where users (or their attributes/avatars) will have multiple identities and identifiers;

- More and more intelligent devices with self-adaptation/context awareness characteristics;

- Service adaptivity, and service configurability, with service platforms providing the agility for ad-hoc coalition of resources.

Chapter Review

1. The definition for network management.

2. What's the difference between TMN management architecture and SNMP-based management architecture? Why the latter surpasses the former from the in practical application?

3. Comparison of CMIP and SNMP.

4. Imagine the possible improvement in SNMP with the evolution of networks?

5. What are the protocols IPFIX, Syslog, and NETCONF used for?

6. Extend the management functions for future networks.

7. The trend of network management.

Chapter 4

Theories and Techniques for Network Management

With the the evolution of networks and the improvement in networks management, researchers and engineers are constantly investigating new theories and techniques to meet the enhancing challenges in network management. A set of enabling technologies are commonly recognized to be potential candidates for network management. Figure 4.1 presents a classification of the existing solutions. Their potential benefits to network management are examined, and their drawbacks and postulates on their prospects are investigated and discussed in this chapter.

4.1 Policy-Based Network Management

4.1.1 Introduction of Policy-Based Management

A policy is the combination of rules and services where rules define the criteria for resource access and usage. A policy is formally defined as an aggregation of policy rules. Each policy rule is composed of a set of conditions and a corresponding set of actions. The condition defines when the policy rule is applicable. Once a policy rule is activated, one or more actions contained by that policy rule may be executed. These actions are associated with either meeting or not meeting the set of conditions specified in the policy rule [CS03].

Policy-based systems have become a promising solution for implementing many forms of large-scale, adaptive systems that dynamically change their behavior in response to changes in the environment or to changing application requirements. This can be achieved by modifying the policy rules interpreted by distributed entities, without recoding or stopping the system. Such dynamic adaptability is fundamentally important in the management of increasingly complex computing systems.

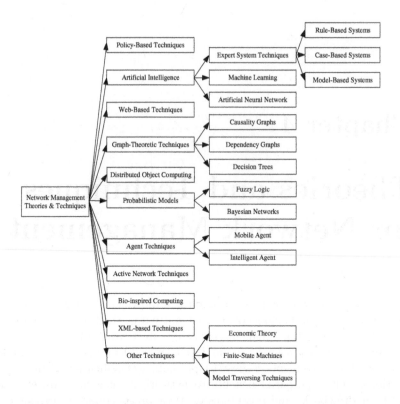

Figure 4.1: Classification of Network Management Techniques

Policy-based management (PBM) is a management paradigm that separates the rules governing the behavior of a system from its functionality. It promises to reduce maintenance costs of information and communication systems while improving flexibility and runtime adaptability. It is today present at the heart of a multitude of management architectures and paradigms, including SLA-driven, business-driven, autonomous, adaptive, and self-* management [BA07].

The policy-based technology could relieve the suffering of managing the large computer systems and free the manager from monitoring the equipments and systems directly and supply a systematic method for establishing, revising, and distributing policies. Policy is a kind of criterion that aims at determining the choice of the actions in an individual system. The criterion is long-lasting, illustrative, and originated from the target of the management.

As a result the policy-based management has the following merits:

- When system requirement alters, it is only necessary to change or add some new policies instead of re-coding.

- To make the best use of the resources by flexible distribution of the resources according to the dynamic information and the different requirements of various service types.

- Different users use different policies, and this is convenient for users and at the same time makes the system more extensible and more maintainable.

- Make the system less dependent on the system manager and make the system more intelligent. Many researchers and organizations have come to do research together on the framework of the policies and its implementation and begin to apply it in the management of the network and wireless network. Indeed the policy-based management is still immature and need to make improvement.

4.1.2 Policy-Based Management Architecture

Several working groups in the IETF and DMTF (Distributed Management Task Force) try to define a standard policy framework and related protocols. See Figure 4.2.

In Figure 4.2, LDAP denotes Lightweight Directory Access Protocol, COPS denotes Common Open Policy Protocol, and CLI denotes Command Line Interface.

It includes the following components:

Policy Management Tool is the server or host where policy management software can do

- policy editing

- policy presentation

- rule translation

- rule validation

- global conflict resolution

Policy Information Repository is a data store for policy information. This data store may be application specific, operating system specific, or an enterprise common repository. For the purpose of this report, the policy information repository is a PBM application specific directory service. Policy information repository can

- store policy information

- search policy information

- retrieve policy information

Policy Decision Point (PDP) is the arbitration component for policy evaluation, which evaluates a state or condition to determine whether a policy enforcement action is required. PDP can work as

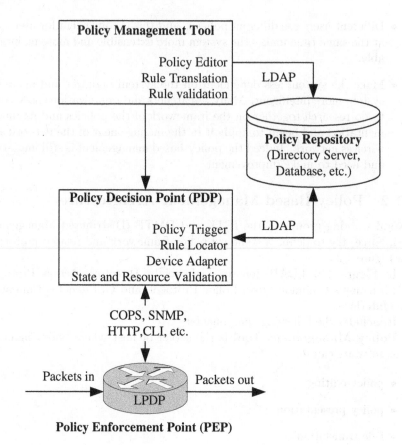

Figure 4.2:   The IETF Policy-Based Management Architecture

- rule locator

- device adapter

- state resource validation (requirements checking)

- policy rule translation

- policy transformation

Policy Enforcement Point (PEP) is a network device, such as a router, switch, firewall or host that enforce the policies received from the PDP. The policies are enforced (through dynamic configuration changes to access control lists, priority queues or other parameters) as directed by the policy decision point. PEP can do

- specified Operation by policy rule

- optional policy rule validation

- feedback

In most implementations of the framework, the Policy Server (Tools), Policy Repository, and PDP are collocated and may potentially be hosted within the same physical device.

Advantages of policy-based management are listed as follows [Str03]:

- providing better-than-best-effort service to certain users

- simplifying device, network, and service management

- requiring less engineers to configure the network

- defining the behavior of a network or distributed system

- managing the increasing complexity of programming devices

- using business requirements and precedures to drive the configuration of the network

4.1.3 Policy-Based Network Management

Policy-Based Network Management is that the network management is accomplished based on policy.

Large-scale networks can now contain millions of components and potentially cross organizational boundaries. Components fail and so other components must adapt to mask these failures. New applications, services, and resources are added or removed from the system dynamically, imposing new requirements on the underlying infrastructure. Users are increasingly mobile, switching between wireless and fixed communication links. To prevent the operators from drowning in excessive detail, the level of abstraction needs to be raised in order to hide system and network specifics. Policies that are derived from the goals of management define the desired behavior of distributed heterogeneous systems and networks and specify means to enforce this behavior. Policy provides a means of specifying and dynamically changing management strategy without coding policy into the implementation. Policy-based management has many benefits of delivering consistent, correct, and understandable network systems. The benefits of policy-based management will grow as network systems become more complex and offer more services (security service and QoS).

Policy-Based Network Management (PBNM) provides a means by which the administration process can be simplified and largely automated [Ver02]. Strassner defined policy-based network management (PBNM) as a way to define business needs and ensure that the network provides the required services [Str03]. In traditional network management approaches, such as SNMP, the usage of network management system has been limited primarily to monitoring status of networks. In PBNM, the information model and policy expressions can be made independent of network management protocols by which they are carried.

The task of managing information technology resources becomes increasingly complex as managers must take heterogeneous systems, different networking technologies, and distributed applications into consideration. As the number of resources to be managed grows, the task of managing these devices and applications depends on numerous system and vendor-specific issues.

Policy-based network management started in the early 1990s [MS93] [KKR95]. Although the idea of policies appeared even earlier, they were used primarily as a representation of information in a specific area of network management: security management [DM89]. The idea of policy comes quite naturally to any large management structure.

In policy-based network management, policies are defined as the rules that govern the states and behaviors of a network. The management system is tasked with:

- the transformation of human-readable specifications of management goals to machine-readable and verifiable rules governing the function and status of a network,

- the translation of such rules to mechanical and device-dependent configurations, and

- the distribution and enforcement of these configurations by management entities.

The most significant benefit of policy-based network management is that it promotes the automation of establishing management-level objectives over a widerange of systems devices. The system administrator would interact with the networks by providing high-level abstract policies. Such policies are device-independent and are stated in a human-friendly manner.

Policy-based network management can adapt rapidly to the changes in management requirements via run-time reconfigurations, rather than reengineer new object modules for deployment. The introduction of new policies does not invalidate the correct operation of a network, provided the newly introduced policies do not conflict with existing policies. In comparison, a newly engineered object module must be tested thoroughly in order to obtain the same assurance.

For large networks with frequent changes in operational directives, policy-based network management offers an attractive solution, which can dynamically translate and update high-level business objectives into executable network configurations. However, one of the key weaknesses in a policy-based network management lies in its functional rigidity. After the development and deployment of a policy-based network management, the service primitives are defined. By altering management policies and modifying constraints, we have a certain degree of flexibility in coping with changing management directives. However, we cannot modify or add new management services to a running system, unlike mobile code or software agents [BX02].

In details, essential characteristics of a PBNM system include:

- Extensibility

  - Customization
  - Expansion
  - Management and provisioning of other services
  - Support extensions through interfaces

- Functionality

  - Configure QoS
  - Ensure bandwidth
  - Control bandwidth
  - Provide application performance analysis
  - Control access
  - Configure usage (authentication and/or encryption)
  - Define QoS treatment of encrypted flows (combine security and QoS Policies)

- Heterogeneity

  - Manage QoS in multi-domain networks
  - Enable end-to-end QoS management
  - Configure security services with gateways from different domains on each side

- Scalability

  - Support hierarchical policy management
  - Enable policy management across multiple policy domains

- Standards Based

  - Support key standards (IETF, ISO, DiffServ, IPSes, etc.) as they are accepted

- Usability

  - Integrate with existing management solutions
  - Hide the detail and present useful concepts and interfaces

PBNM can be used in dealing with complex network management tasks. Figure 4.3 [KTT+05] illustrates a PBNM automated network device configuration / change management capability in an integrated environment.

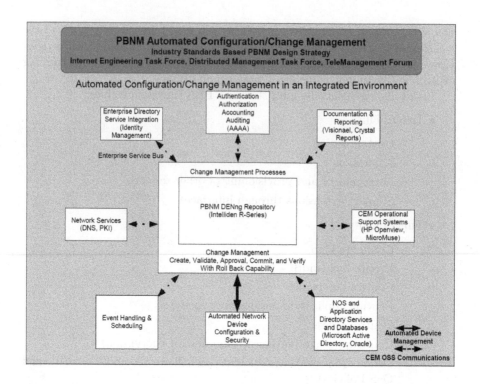

Figure 4.3: PBNM Automated Configuration/Change Management in an Integrated Environment

4.2 Artificial Intelligence Techniques for Network Management

For network management, it is very difficult to find a network manager for each management center who has entire knowledge of the network. It is also important that the network managers be available round the clock to handle the network and to make sure that it runs in a healthy condition. Having such manpower all the time and working in the event of network failure, when everybody expects the network to come back to normalcy in the stipulated time, is a very tough job. This demands the use of an automated program, which either imitates the manager in his/her absence or assists in the presence of the manager.

Automation of network management activities can benefit from the use of Artificial Intelligence (AI) technologies, including fault management, performance analysis, and traffic management.

AI includes such automated programs in the form of an Expert System (ES), which are expected to mimic an expert of one particular domain. An idea to deploy such expert systems to assist the network managers motivates the use of artificial intelligence methods to network management [KV97].

4.2.1 Expert Systems Techniques

The term "expert system" refers to a system that uses contemporary technology to store and interpret the knowledge and experience of a human expert, sometimes several experts, in a specific area of interest. By accessing this computer-based knowledge, an individual is able to get the benefit of "expert advice" in the particular area [LG91].

Figure 4.4 is a high-level diagram of the components of an expert system.

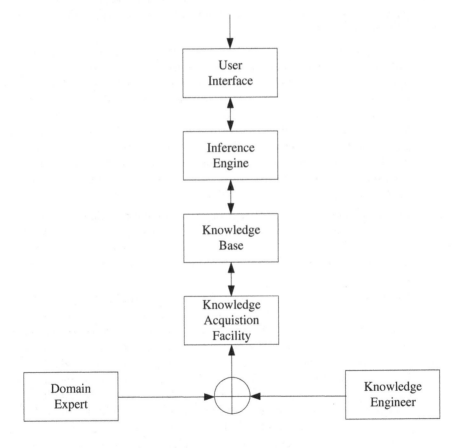

Figure 4.4: Expert System Functional Diagram

Expert systems try to reflect actions of a human expert when solving problems in a particular domain. The knowledge base is where the knowledge of human experts in a specific field or task is represented and stored. It contains a set of rules or cases with the knowledge about a specific task, that is, an instance of that class of problems. The inference engine of an expert system contains the strategy to solve a given class of problems using, e.g., the rules in a particular sequence. It is usually set up to mimic the reasoning or problem-solving ability that the human expert would use to arrive at a conclusion. Rule-based, case-based, and model-based approaches are popular expert systems that are widely used in fault management [PMM89].

- **Rule-based systems**

  Most expert systems use rule-based representation of their knowledge base. In the rule-based approach, the general knowledge of a certain area is contained in a set of rules and the specific knowledge, relevant for a particular situation, is constituted of facts, expressed through assertions and stored in a database.

  There are two operation modes in a rule-based system. One is the forward mode, which departs from an initial state and constructs a sequence of steps that leads to the solution of the problem ("goal"). When it comes to a fault diagnosis system, the rules would be applied to a database containing all the alarms received, until a termination condition involving one fault is reached. The other is the backward mode, which starts from a configuration corresponding to the solution of the problem and constructs a sequence of steps that leads to a configuration corresponding to the initial state. The same set of rules may be used for the two operation modes [Ric83].

  In the domain of fault localization, the inference engine usually uses a forward-chaining inference mechanism, which operates in a sequence of rule-firing cycles. In each cycle the system chooses rules for execution whose antecedents (conditions) match the content of the working memory.

  Expert systems applied to the fault localization problem differ with respect to the structure of the knowledge they use. The approaches that rely solely on surface knowledge are referred to as rule-based reasoning systems. The research on rule-based fault localization systems addresses the structure of the knowledge base and the design of the rule-defined language. [Lor93] organizes the system of rules by distinguishing between core and customized knowledge. The core knowledge may be understood as a generic or reusable knowledge. It is useful for identifying an approximate location of a fault in a large network.

  Rule-based systems, which rely solely on surface knowledge, do not require profound understanding of the underlying system architectural and operational principles and, for small systems, may provide a powerful tool for eliminating the least likely hypotheses [Kat96].

  However, rule-based systems possess a number of disadvantages that limit their usability for fault isolation in more complex systems. The downside of rule-based systems include the inability to learn from experience, inability to deal with unseen problems, and difficulty in updating the system knowledge [Lew93]. Rule-based systems are difficult to maintain because the rules frequently contain hard-coded network configuration information. Although approaches have been proposed to automatically derive correlation rules based on the observation of statistical data [KMT99], it is still necessary to regenerate a large portion of correlation rules when the system configuration changes. Rule-based systems are also inefficient and unable to deal with inaccurate information [SMA01]. The lack of structure in the system of rules typically makes it very difficult to allow the reusability of rules that seems so intuitive in hierarchically built networks. Another problem is that rule-based systems get convoluted if timing constraints are included in the reasoning process. Also, rule interactions may result in unwanted sideeffects, and are difficult to verify and change [WBE+98]. [KK98] extends a production rule interpreter and to enable a systematic prediction of the effects of policy executions and to allow for a better impact analysis in case of policy changes.

The rule-based systems rely heavily on the expertise of the system manager. The rules depend on the prior knowledge about the fault conditions on the network, and do not adapt well to the evolving network environment [FKW96]. Thus it is possible that entirely new faults may escape detection. Furthermore, even for a stable network there are no guarantees that an exhaustive database has been created.

[WEW99] introduced one system called ANSWER (Automatic Network Surveillance with Expert Rules), an expert system used in monitoring and maintaining the 4ESS switches in the AT&T long distance network. The knowledge base uses object-oriented programming (C++) to model the 4ESS switch leading to flexible and efficient design, i.e., only components with abnormal activities get instantiated, thus realizing reductions in time and space. The knowledge base interacts with the actual switch in two ways: 1) it receives events as input; and 2) it issues commands (e.g., to request diagnostics to be run) as output. The rule-based component of the system has been implemented using an extension to the C++ language called R++. This has led to the tight integration of the knowledge base with the rest of the system.

[She96] presents the HP OpenView Event Correlation Service (ECS). This package is a commercial product integrated into the HP OpenView Distributed Management (DM) postmaster. It consists of two components: the ECS designer and the ECS engine. The ECS Designer is a graphical user interface (GUI) development environment where correlation rules can be developed interactively by selecting, connection, and configuring logical processing blocks (nodes). The process of combining different nodes creates a correlation circuit where events flow from a source node through the path(s) of the defined circuit and exit through a sink node. The ECS engine is a run-time correlation engine. It executes a set of download correlation rules that control the processing of event streams.

- **Case-based systems**

  As an alternative to the rule-based approach, some researchers propose a technique-denominated case-based reasoning (CBR) [Sla91] [WTM95] [MT99]. Here, the basic unit of knowledge is not a rule but a case. Cases contain registers with the most relevant characteristics of past episodes and are stored, retrieved, adapted, and utilized in the solution of new problems. The experience obtained from the solution to these new problems constitutes new cases, which are added to the database for future use. Thus, the system is able to acquire knowledge through its own means, and it is not necessary to interview human experts. Another relevant characteristic of the CBR systems is their ability to modify their future behavior according to the current mistakes. In addition, a case-based system may build solutions to the unheard-of problems through the adaptation of past cases to the new situations.

  Since the development of CBR systems started in the 1980s, several challenges have stimulated the researchers' creativity: how to represent the cases; how to index them to allow their retrieval when necessary; how to adapt an old case to a new situation to generate an original solution; how to test a proposed solution and identify it as either a success or a failure; and how to explain and repair the fault of a suggested solution to originate a new proposal.

  The problem of case adaptation is studied in [LD93]. It described a technique named parameterized adaptation, which is based on the existence in a trouble ticket, of a certain relationship among the variables that describe a problem and the variables that specify the corresponding solution. A CBR system takes into account the parameters of this relationship in the proposition of a solution for the case under analysis. To represent the parameters, the use of linguistic variables (i.e., the ones that assume linguistic values, instead of numeric values) and the provision of functions is proposed, so that the parameters' numeric values are translated into grades of membership in a fuzzy set.

  To store and retrieve the knowledge on the solution of past problems, [DV95] presents the concept of master ticket, which contains a generalization of information on the faults instead of information on a single fault (case). Thus, when a master ticket is retrieved, it must be instantiated before the information it contains may be applied to a particular case. The goal of this procedure is to facilitate the access to the information and the involved node's addresses. To instantiate a master ticket consists of substituting its parameters for the real values of the case under consideration.

- **Model-based systems**

  Model-based reasoning (MBR) is a paradigm that arose from artificial intelligence, which has several applications in alarm correlation. MBR represents a system through a structural model and a functional model, while the rules are based on empirical associations in traditional rule-based systems. In a management system for a network, the structural representation includes a description of the network elements and of the topology (i.e., connectivity and containment). The representation of functional behavior describes the processes of event propagation and event correlation [JW95].

  Model-based reasoning systems play an important role in the device-level fault management and in the execution escalation rules or fault management policies such as which trouble tickets have to be created under which circumstances. In this role the expert system serves as an integrator between the different fault management techniques.

  In model-based expert systems, conditions associated with the rules usually include predicates referring to the system model. The predicates test the existence of a relationship among system components. The model is usually defined by an object-oriented paradigm [DSW+91] [FJM+98] [JW93], and it frequently has the form of a graph of dependencies among system components. A different model is proposed in SINERGIA [BBM+93], which represents structural knowledge as a set of network topology templates selected in such a way that any network topology may be expressed as instances of these templates. All possible alarm patterns for each template are listed along with fault diagnosis functions.

Currently, many network management systems employing AI technologies for fault diagnosis, however, they have their limitations [GKO+96]:

- Expert Systems (ESs) cannot handle new and changing data. Rules are brittle and not robust when faced with unforeseen situations (e.g., a new combination of alarms due to changing network topology).

- They cannot learn from experience (i.e., they cannot use analogy to reason from past experiences or remember past successes and failures in the context of a current problem). The rules that are incorporated at development time cannot easily adapt as the network evolves.

- They do not scale well to large dynamic real-world domains. It is difficult, especially for technicians or operators not familiar with AI, to add new rules without a comprehensive understanding of what the current rule base is and how a new rule may impact the rule base.

- They require extensive maintenance when the domain knowledge changes; new rules have to be added and old rules adapted or deleted.

- They are not good at handling probability or uncertainty. Fuzzy logic can be employed to create fuzzy rules. However, fuzzy expert systems still have the problems discussed above.

- They have difficulty in analyzing large amounts of uncorrelated, ambiguous, and incomplete data. The domain must be well understood and thought out. This is not entirely possible in domains such as fault management.

These drawbacks argue for the use of different AI technologies that can overcome the above-mentioned difficulties, either alone or as an enhancement of ESs. Probabilistic methods are appropriate for correlation, while symbolic methods such as case-based reasoning (CBR) or expert systems are appropriate for fault identification. In many cases it is beneficial to use these technologies in cooperation with each other.

4.2.2 Machine Learning Techniques

Machine learning is a subfield of artificial intelligence that is concerned with the design and development of algorithms and techniques that allow computers to "learn." In general, there are two types of learning: inductive and deductive. Inductive machine learning methods extract rules and patterns out of massive data sets.

The major focus of machine learning research is to extract information from data automatically, by computational and statistical methods. Hence, machine learning is closely related not only to data mining and statistics but also to theoretical computer science.

Applications for machine learning include natural language processing, syntactic pattern recognition, search engines, medical diagnosis, bioinformatics, brain-machine interfaces and cheminformatics, detecting credit card fraud, stock market analysis, classifying DNA sequences, speech and handwriting recognition, object recognition in computer vision, game playing, and robot locomotion. In network management, machine learning can play an important role in network fault management, self-configuration, and optimization.

Some machine learning systems attempt to eliminate the need for human intuition in data analysis, while others adopt a collaborative approach between human and machine. But human intuition cannot be entirely eliminated, since the system's designer must specify how the data are to be represented and what mechanisms will be used to search for a characterization of the data. Machine learning can be viewed as an attempt to automate parts of the scientific method.

Machine learning algorithms are organized into a taxonomy, based on the desired outcome of the algorithm. Common algorithm types include:

- Supervised learning: in which the algorithm generates a function that maps inputs to desired outputs. One standard formulation of the supervised learning task is the classification problem: the learner is required to learn (to approximate) the behavior of a function that maps a vector $[X_1, X_2, ..., X_n]$ into one of several classes by looking at several input-output examples of the function.

- Unsupervised learning: an agent that models a set of inputs; labeled examples are not available.

- Semi-supervised learning: which combines both labeled and unlabeled examples to generate an appropriate function or classifier.

- Reinforcement learning: in which the algorithm learns a policy of how to act given an observation of the world. Every action has some impact in the environment, and the environment provides feedback that guides the learning algorithm.

- Transduction: similar to supervised learning but does not explicitly construct a function; instead, tries to predict new outputs based on training inputs, training outputs, and test inputs which are available while training.

- Learning to learn: in which the algorithm learns its own inductive bias based on previous experience.

Problem formulations in machine learning [DL07] can be classified as:

- Learning for classification and regression: Classification involves assigning a test case to one of a finite set of classes, whereas regression instead predicts the case's value on some continuous variables or attribute. In the context of network diagnosis, one classification problem is deciding whether a connection failure is due to the target site being down, the target site being overloaded, or the ISP service being down. An analogous regression problem might involve predicting the time it will take for the connection to return. Cases are typically described as a set of values for discrete or continuous attributes or variables. For example, a description of the network's state might include attributes for packet loss, transfer time, and connectivity. Some work on classification and regression instead operates over relational descriptors. Thus, one might describe a particular situation in terms of node connections and whether numeric attributes at one node (e.g., buffer utilization) are higher than those at an adjacent node.

- Learning for acting and planning: Action selection can occur in a purely reactive way, ignoring any information about past actions. This version has a straightforward mapping onto classification, with alternative actions corresponding to distinct classes from which the agent can choose based on descriptions of the world state. One can also map it onto regression, with the agent predicting the overall value or utility of each action in a given world state.

Both approaches can also be utilized for problem solving, planning, and scheduling. These involve making cognitive choices about future actions, rather than about immediate actions in the environment. Such activities typically involve search through a space of alternatives, which knowledge can be used to constrain or direct. This knowledge may take the form of classifiers for which action to select or regression functions over actions or states. However, it can also be cast as larger-scale structures called macro-operators that specify multiple actions that should be carried out together.

For example, one might implement an interactive tool for network configuration that proposes, one step at a time, a few alternative components to incorporate or connections among them. The human user could select from among these recommendations or reject them all and select another option. Each such interaction would generate a training instance for use in learning how to configure a network, which would then be used on future interactions. One can imagine similar adaptive interfaces for network diagnosis and repair.

- Learning for interpretation and understanding: One can state a number of different learning tasks within the explanatory framework. The most tractable problem assumes that each training case comes with an associated explanation cast in terms of domain knowledge. This formulation is used commonly within the natural language community, where the advent of "tree banks" has made available large corpora of sentences with their associated parse trees. The learning task involves generalizing over the training instances to produce a model that can be used to interpret or explain future test cases. Naturally, this approach places a burden on the developer, since it requires hand construction of explanations for each training case, but it greatly constrains the learning process, as it effectively decomposes the task into a set of separate classification or density estimation tasks, one for each component of the domain knowledge.

Table 4.1 summarizes the main formulations in machine learning.

Machine Learning for Network Management

1. Anomaly detection and fault diagnosis

There are several issues that arise in anomaly detection.

First, one must choose the level of analysis and the variables to monitor for anomalies. This may involve first applying methods for interpreting

Table 4.1: Summary of Machine Learning Problem Formulations

Formulation	Performance Task
Classification & Regression	predict y given x
	predict rest of x given part of x
	predict $P(x)$ given x
Acting & Planning	iteratively choose action a in state s
	choose actions $< a_1, ..., a_n >$ to achieve goal g
	find setting s to optimize objective $J(s)$
Interpretation & Understanding	parse data stream into tree structure of objects
	or events

and summarizing sensor data. In the Knowledge Plane, one can imagine having whole hierarchies of anomaly detectors looking for changes in the type of network traffic (e.g., by protocol type), in routing, in traffic delays, in packet losses, in transmission errors, and so on. Anomalies may be undetectable at one level of abstraction but easy to detect at a different level. For example, a worm might escape detection at the level of a single host, but be detectable when observations from several hosts are combined.

The second issue is the problem of false alarms and repeated alarms. Certain kinds of anomalies may be unimportant, so network managers need ways of training the system to filter them out. Supervised learning methods could be applied to this problem.

Fault isolation requires the manager to identify the locus of an anomaly or fault within the network. For example, if a certain route has an especially heavy load, this may be due to changes at a single site along that route rather than to others. Hence, whereas anomaly detection can be performed locally (e.g., at each router), fault isolation requires the more global capabilities of the Knowledge Plane to determine the scope and extent of the anomaly.

The activity of diagnosis involves drawing some conclusions about the cause of the anomalous behavior. Typically, this follows fault isolation, although in principle one might infer the presence of a specific problem without knowing its precise location. Diagnosis may involve the recognition of some known problems, say, one the network manager has encountered before, or the characterization of a new problem that may involve familiar components.

One can apply supervised learning methods to let a network manager teach the system how to recognize known problems.

2. Intrusion detection

Responding to intruders (human, artificial, or their combination) and keeping networks and applications safe encompass a collection of tasks that are best explained depending on the time at which the network manager performs them.

Prevention Tasks. Network managers try to minimize the likeliness of future intrusions by constantly auditing the system and eliminating threats beforehand. A network manager proactively performs security audits testing the computer systems for weaknesses, vulnerabilities, or exposures. However, scan tools (e.g., Nessus, Satan, and Oval) used for penetration or vulnerability testing only recognize a limited number of vulnerabilities given the ever increasing frequency of newly detected possibilities for breaking into a computer system or disturbing its normal operation. Thus, network managers continually update scan tools with new plug-ins that permit them to measure new vulnerabilities. Once the existence of a vulnerability or exposure is perceived, network managers assess the convenience of discontinuing the service or application affected until the corresponding patch or intrusion detection signature is available. A trade-off between risk level and service level is made in every assessment.

Network managers aim at shrinking the window of vulnerability, the time gap between when a new vulnerability or exposure is discovered and a preventative solution (patch, new configuration, etc.) is provided. A basic strategy to accomplish that objective is based on two conservative tasks: first, minimizing the number of exposures (i.e., disable unnecessary or optional services by configuring firewalls to allow only the use of ports that are necessary for the site to function) and, second, increasing awareness of new vulnerabilities and exposures (e.g., the subscription model that Partridge discusses with relation to worms).

Finally, network managers continuously monitor the system so that pre-intrusion behavioral patterns can be understood and used for further reference when an intrusion occurs. Monitoring is an ongoing, preventive task.

Detection Tasks. The sooner an intrusion is detected, the more chances there are for impeding an unauthorized use or misuse of the computer system. Network managers monitor computer activities at different levels of detail: system call traces, operating system logs, audit trail records, resource usage, network connections, etc. They constantly try to fuse and correlate real-time reports and alerts stemming from different security devices (e.g., firewalls and intrusion detection systems) to stop suspicious activity before it has a negative impact (i.e., degrading or disrupting operations). Different sources of evidence are valuable given the evolving capabilities of intruders to elude security devices. The degree of suspicion and malignancy associated with each report or alert still requires continuous human oversight. Consequently, network managers are continually overwhelmed with a vast amount of log information and bombarded with countless alerts. To deal with this onslaught, network managers often tune security devices to reduce the number of false alerts even though this increases the risk of not detecting real intrusions.

The time at which an intrusion is detected directly affects the level of damage that an intrusion causes. An objective of network managers is to reduce the window of penetrability, the time span that starts when a computer system has been broken into and extends until the damage has been completely repaired. The correct diagnosis of an intrusion allows a network manager to initiate the most convenient response. However, a trade-off between quality and rapidness is made in every diagnostic.

Response and Recovery Tasks. As soon as a diagnostic on an intrusion is available, network managers initiate a considered response. This response tries to minimize the impact on the operations (e.g., do not close all ports in a firewall if only blocking one IP address is enough). Network managers try to narrow the window of compromisibility of each intrusion. The time gap that starts when an intrusion has been detected and ends when the proper response has taken effect deploying automatic intrusion response systems. Nevertheless, these systems are still at an early stage and even fail at providing assistance in manual responses. Therefore, network managers employ a collection of ad-hoc operating procedures that indicate how to respond and recover from a type of intrusion. The responses to an attack range from terminating a user job or suspending a session to blocking an IP address or disconnecting from the network to disable the compromised service or host. Damage recovery or repairing often requires maintaining the level of service while the system is being repaired, which makes this process difficult to automate.

Once the system in completely recovered from an intrusion, network managers collect all possible data to thoroughly analyze the intrusion, trace back what happened, and evaluate the damage. Thus, system logs are continuously backed up. The goal of postmortem analysis is twofold. On the one hand, it gathers forensic evidence (contemplating different legal requirements) that will support legal investigations and prosecution and, on the other hand, it compiles experience and provides documentation and procedures that will facilitate the recognition and repelling of similar intrusions in the future.

Ideally, the ultimate goal of a network manager is to make the three windows (vulnerability, penetrability, and compromisibility) of each possible intrusion converge into a single point in time. Tasks for responding to intruders (human, artificial, or a combination of both) should not differ significantly from those tasks needed to recover from non-malicious errors or failures.

3. **Network configuration and optimization**

The problem of the design and configuration of engineered systems has been studied in artificial intelligence since the earliest days. Configuration is generally defined as a form of routine design from a given set of components or types of components (i.e., as opposed to designing the components themselves).

The simplest task is parameter selection, where values are chosen for a set of global parameters in order to optimize some global objective function. Two classic examples are the task of setting the temperature, cycle time, pressure, and input/output flows of a chemical reactor and the task of controlling the rate of cars entering a freeway and the direction of flow of the express lanes. If a model of the system is known, this becomes purely an optimization problem, and many algorithms have been developed in operations research, numerical analysis, and computer science to solve such problems.

The second task is compatible parameter selection. Here, the system consists of a set of components that interact with one another to achieve overall system function according to a fixed topology of connections. The effectiveness of the interactions is influenced by parameter settings, which must be compatible in order for sets of components to interact. For example, a set of hosts on a subnet must agree on the network addresses and subnet mask in order to communicate using IP. Global system performance can depend in complex ways on local configuration parameters. Of course, there may also be global parameters to select as well, such as the protocol family to use.

The third task is topological configuration. Here, the system consists of a set of components, but the topology must be determined. For example, given a set of hosts, gateways, file servers, printers, and backup devices, how should the network be configured to optimize overall performance? Of course, each proposed topology must be optimized through compatible parameter selection.

Finally, the most general task is component selection and configuration. Initially, the configuration engine is given a catalog of available types of components (typically along with prices), and it must choose the types and quantities of components to create the network (and then, of course, solve the Topological Configuration problem of arranging these components).

A different challenge arises when attempting to change the configuration of an existing network, especially if the goal is to move to the new configuration without significant service interruptions. Most configuration steps require first determining the current network configuration, and then planning a sequence of reconfiguration actions and tests to move the system to its new configuration. Some steps may cause network partitions that prevent further (remote) configuration. Some steps must be performed without knowing the current configuration (e.g., because there is already a network partition, congestion problem, or attack).

Parameter Selection. As we discussed above, parameter selection becomes optimization (possibly difficult, non-linear optimization) if the model of the system is known. Statisticians have studied the problem of empirical optimization in which no system model is available.

Compatible Parameter Configuration. The standard AI model of compatible parameter configuration is known as the constraint satisfaction problem (CSP). This consists of a graph where each vertex is a variable that can take values from set of possible values and each edge encodes a pair-wise constraint between the values of the variables that it joins. A large family of algorithms have been developed for finding solutions to CSPs efficiently. In addition, it is possible to convert CSPs into Boolean satisfiability problems, and very successful randomized search algorithms, such as WalkSAT, have been developed to solve these problems. The standard CSP has a fixed graph structure, but this can be extended to include a space of possible graphs and to permit continuous (e.g., linear algebraic) constraints.

In both refinement and repair-based methods, constraint satisfaction methods are typically applied to determine good parameter values for the current proposed configuration. If no satisfactory parameter values can be found, then a proposed refinement or repair cannot be applied, and some other refinement or repair operator must be tried. It is possible for the process to reach a dead end, which requires backtracking to some previous point or restarting the search. Component Selection and Configuration. The refinement and repair-based methods described above can also be extended to handle component selection and configuration. Indeed, our local network configuration example shows how refinement rules can propose components to include in the configuration. Similar effects can be produced by repair operators.

Changing Operating Conditions. The methods discussed so far only deal with the problem of optimizing a configuration under fixed operating conditions. However, in many applications, including networking, the optimal configuration may need to change as a result of changes in the mix of traffic and the set of components in the network. This raises the issue of how data points collected under one operating condition (e.g., one traffic mix) and be used to help optimize performance under a different operating condition. To our knowledge, there is no research on this question.

Artificial Neural Networks for Network Management

An Artificial Neural Network (ANN) is a system constituted of elements (neurons) interconnected according to a model that tries to reproduce the functioning of the neural network existing in the human brain. Conceptually, each neuron may be considered as an autonomous processing unit, provided with local memory and with unidirectional channels for the communication with other neurons. The functioning of an input channel in an ANN is inspired by the operation of a dendrite in biological neurons. In an analog way, an output channel has an axon as its model. A neuron has only one axon, but it may have an arbitrary number of dendrites (in a biological neuron there are around ten thousand dendrites). The output "signal" of a neuron may be utilized as the input for an arbitrary number of neurons [MK94b] [Mea89].

In its simplest form, the processing carried out in a neuron consists of affecting the weighted sum of the signals present in their inputs and of generating an output signal if the result of the sum surpasses a certain threshold. In the most general case, the processing may include any type of mathematical operation on the input signals, also taking into consideration the values stored in the neuron's local memory.

One of the main motivations for the development of the ANN is the utilization of computers to deal with a class of problems that are easily solved by the human brain, but which are not of effectively treated with the exclusive utilization of conventional programming paradigms.

The distributed control and storage of data and parallelism are remarkable features of the ANN. Besides that, an ANN does not require previous knowledge of the mathematical relationship between inputs and outputs, which may be automatically learned, during the system's normal operation. This makes them, at first, a good alternative for applications (such as alarm correlation and fault diagnosis) where the relationships between faults and alarms are not always well defined or understood and where the available data is sometimes ambiguous or inconsistent [CMP89].

Feedforward neural networks have already been proven effective in medical diagnosis, target tracking from multiple sensors, and image/data compression. It is therefore plausible that ANNs would be effective for the similar problem of alarm correlation, found in fault diagnosis. In fact, the following properties of multilayer feedforward neural networks make them a powerful tool for solving these problems.

- ANN can recognize conditions similar to previous conditions for which the solution is known (i.e., pattern matching).

- They can approximate any function, given enough neurons, including boolean functions and classifiers. This gives ANN great flexibility in being able to be trained for different alarm patterns.

- They can generalize well and learn an approximation of a given function, without requiring a deep understanding of the knowledge domain. This is especially important in new technological areas such as ATM switch networks.

- They provide a fast and efficient method for analyzing incoming alarms.

- They can handle incomplete, ambiguous, and imperfect data.

In a feedforward neural net, shown in Figure 4.5, the neurons are arranged into layers, with the outputs of each layer feeding into the next layer. This model has a single input layer, a single output layer, and zero, one, or more hidden layers. As the name suggests, all connections are in the forward direction where there is no feedback. Feedforward networks are useful because of their ability to approximate any function, given enough neurons, and their ability to learn (generalize) from samples of input-output pairs. Learning is accomplished by adjusting the connection weights in response to input-output pairs, and training can be done either off-line, or on-line during actual use. Depending on how the training is done, these ANNs can be characterized as being trained by supervised methods or by unsupervised methods.

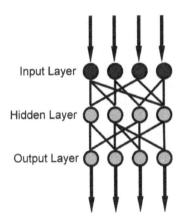

Figure 4.5: Model of a Feedforward Neural Network

Supervised ANNs training data consist of correct input vector/output vector pairs as examples, used to adjust the neural net connection weights. An input vector is applied to the NN, the output vector obtained from the NN is compared with the correct output vector, and the connection weights are changed to minimize the difference. A well-trained neural net can successfully generalize what it has learned from the training set (i.e., given an input vector not in the training set, it produces the correct output vector most of the time).

In unsupervised training there is no training data based on known input/output pairs. The NN discovers patterns, regularities, correlations, or categories in the input data and accounts for them in the output. For example, an unsupervised neural net where the variance of the output is minimized could serve as a categorizer that clusters inputs into various groups. Unsupervised training is typically faster than supervised training and provides the opportunity to present patterns to operations personnel who can identify new output relations. For these reasons unsupervised training is used even in situations where supervised training is possible. However, for the domain of alarm correlation, input-output pairs can be easily produced, making supervised trained ANN a plausible choice for alarm correlation.

Some research uses the ability of neural networks to predict future behavior of general nonlinear dynamic systems. In this approach, a neural network predicts normal system behavior based on past observations and the current state of the system. A residual signal is generated based on a comparison between the actual and predicted behavior, and a second neural network is trained to detect and classify the alarms based on characteristics of the residual signal. This method can be used to identify basic categories for the alarms.

An additional approach is to cast the pattern recognition problem into an optimization problem, making Hopfield ANN an appropriate tool for alarm correlation. This type of NN operates by using gradient methods to find a local minimum of a quadratic energy function that represents an optimization problem, and whose coefficients depend on the network's interconnection strengths. Methods such as mean-field annealing, repeated trials with random initial states, and tabu learning can be used to find a local minimum that is nearly optimal. For example, in alarm correlation, the optimization problem is to identify the hypothesis that best explains the observed data.

ANN is appropriate in fault management since it can analyze patterns of common behavior over circuits, and/or can handle ambiguity, and incomplete data. For example, ANNs have been used for the purpose of fault localization [GH98] [Wie02]. The disadvantage of neural network systems is that they require long training periods [GH96], and that their behavior outside their area of training is difficult to predict [WBE+98].

For AI in network management, there are still some challenges for research [DL07], such as:

- From supervised to autonomous learning

- From off-line to one-line learning

- From fixed to changing environments

- From centralized to distributed learning

- From engineered to constructed representations

- From knowledge-learn to knowledge-rich learning

- From direct to declarative models.

4.3 Graph-Theoretic Techniques for Network Management

Graph-theoretic techniques rely on a graph model of the system, called a Fault Propagation Model (FPM), which describes which symptoms may be observed if a specific fault occurs [KP97]. The FPM represents all the faults and symptoms that occur in the system. The observed symptoms are mapped into the FPM nodes. The fault localization algorithm analyzes the FPM to identify the best explanation for the observed symptoms.

Graph-theoretic techniques require the priori specification of how a failure condition or alarm in one component is related to failure conditions or alarms in other components [HCF95]. The creation of the FPM requires an accurate knowledge of current dependencies among abstract and physical system components. The efficiency and accuracy of the fault localization algorithm depends on the accuracy of this a priori specification. The FPM takes the form of causality or dependency graph.

4.3.1 Causality Graph Model

Network management consists mainly of monitoring, interpreting, and handling events. In management an event is defined as an exceptional condition in the operation of networks. Some problems are directly observable, while others can only be observed indirectly from their symptoms. Symptoms are defined as the observable events. However, a symptom cannot be directly handled; instead its root cause problem needs to be handled to make it go away. Relationships are essential components of the correlation, because problems and symptoms propagate from one object to another relationship.

A natural candidate for representing the problem domain is the causality graph model [YKM+96]. Causality induces a partial order relationship between events.

A causality graph is a directed acyclic graph $G_c(E, C)$ whose nodes E correspond to events and whose edges C describe cause-effect relationships between events. An edge $(e_i, e_j) \in C$ represents the fact that event e_i causes event e_j, which is denoted as $e_i \rightarrow e_j$ [HSV99]. The nodes of a causality graph may be marked as problems or symptoms. Some nodes are neither problems nor symptoms, while others may be marked as problems and symptoms at the same time. Causality graph edges may be labeled with a probability of the causal implication. Similarly, it is possible to assign a probability of independent occurrence to all nodes labelled as problems.

For example, Figure 4.6(a) depicts the causality graph of a network consisting of 7 nodes. Certain symptoms are directly caused by particular problem or indirectly by other symptoms; for instance, symptom 7 in Figure 4.6(a) is generated by node 1 or by node 6.

To proceed with correlation analysis, it is necessary to identify the nodes in the causality graph corresponding to symptoms and those corresponding to

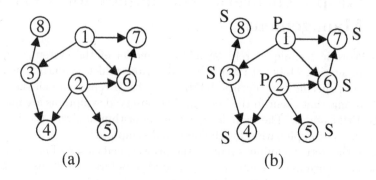

(a)   (b)

Figure 4.6: (a) Causality Graph. (b) Labeling of the Graph.

problems. A problem is an event that may require handling while a symptom (alarm) is an event that is observable. Nodes of a causality graph may be marked as problems (P) or symptoms (S) as in Figure 4.6(b).

4.3.2 Dependency Graph Model

A dependency graph is a directed graph $G = (O, D)$, where O is a finite, non-empty set of objects and D is a set of edges between the objects. Each object may be associated with a probability of its failure independent of other objects. The directed edge $(o_i, o_j) \in D$ denotes the fact that an error or fault in o_i may cause an error in o_j. Every directed edge is labeled with a conditional probability that the object at the end of an edge fails, provided that the object at the beginning of an edge fails [Kat96] [KS95]. Note that the presented dependency graph models all possible dependencies between managed objects in networks. In reality the graph could be reduced based on information on current open connections.

Many approaches using dependency graph models assume that an object may fail in only one way. If this is the case in a real system, then a failure of an object may be represented as an event. In this case, the two representations, causality graph and dependency graph, are equivalent. When multiple failures may be associated with a single object, they can typically be enumerated into a small set of failure modes, such as complete failure, abnormal transmission delay, high packet loss rate and so on. The dependency graph then associates multiple failure modes with each object, whereas dependency edges between objects are weighted with probability matrices rather than with single probability values, where each matrix cell indicates the probability with which a particular failure of an antecedent object causes a particular failure of a dependent object [SS02]. The dependency graph may still be mapped into a causality graph by creating a separate causality graph node for each object and each of its failure modes, and then connecting the nodes accordingly.

While the use of a dependency graph as a system model has many advantages (e.g., it is more natural and easier to build), a causality graph is more suitable for the task of fault localization as it provides a more detailed view of the system and is able to deal with a simple notion of an event rather than potential multi-state system objects.

A dependency graph provides generality because different types of networks can be easily modeled. It is flexible in the sense that we can easily modify the representation as the active picture of the observed system changes. It is simple because it gives a manageable model and it has the necessary complexity to represent real systems. It also has the property of similarity, which means that similar systems have similar representations. For example, if we add another entity in the graph, the representation changes only slightly.

Most graph-theoretic techniques reported in the literature allow the FPM to be nondeterministic by modeling prior and conditional failure probabilities. However, many of these techniques require the probability models to be restricted to canonical models such as OR and AND models [KS95] [KYY+95] [SS02]. An OR model combines possible causes of an event using logical operator OR, meaning that at least one of the possible causes has to exist for the considered event to occur. An AND model uses the logical operator AND, instead. Some techniques may be extended to work with hybrid models that allow the relationship between an event and its possible causes to have the form of an arbitrary logical expression.

4.3.3 Decision Trees

Decision tree is a graphical representation, in which all possible outcomes and the paths may be reached. Decision tree is often used in classification tasks.

In decision tree, the top layer consists of input nodes (e.g., status observations in networks). Decision nodes determine the order of progression through the graph. The leaves of the tree are all possible outcomes or classifications, while the root is the final outcome (for example, a fault prediction or detection). Nearly all expert systems can most appropriately be diagrammed as a decision tree. The root of the tree represents the first test, while the leaves (nodes that do not lead to further nodes) represent the set of possible conclusions or classifications.

Three popular rules are applied in the automatic creation of classification trees. The Gini rule splits off a single group of as large a size as possible, whereas the entropy and twoing rules find multiple groups comprising as close to half the samples as possible. Both algorithms proceed recursively down the tree until stopping criteria are met.

The Gini rule is typically used by programs that build ("induce") decision trees using the CART algorithm. Entropy (or information gain) is used by programs that are based on the C4.5 algorithm [Qui93]. A brief comparison of these two criterion can be seen under decision tree formulae.

More information on automatically building ("inducing") decision trees can be found under Decision tree learning.

Among decision support tools, decision trees (and influence diagrams) have several advantages:

- are simple to understand and interpret. People are able to understand decision tree models after a brief explanation.

- have value even with little hard data. Important insights can be generated based on experts describing a situation (its alternatives, probabilities, and costs) and their preferences for outcomes.

- use a white box model. If a given result is provided by a model, the explanation for the result is easily replicated by simple math.

- can be combined with other decision techniques.

- can be used to optimize combination problems in complex systems.

One area of application for decision trees is systematically listing a variety of functions. The simplest general class of functions to list is the entire set n^k. We can create a typical element in the list by choosing an element of n and writing it down, choosing another element (possibly the same as before) of n and writing it down next, and so on until we have made k decisions. This generates a function in one line form sequentially: First $f(1)$ is chosen, then $f(2)$ is chosen and so on. We can represent all possible decisions pictorially by writing down the decisions made so far and then some downward "edges" indicating the possible choices for the next decision.

Here is an example of a decision tree for the functions 2^3, see Figure 4.7.

The set $V = \{R, 1, 2, 11, 12, 21, 22, 111, 112, 121, 122, 211, 212, 221, 222\}$ is called the set of vertices of the decision tree. The vertex set for a decision tree can be any set, but must be specified in describing the tree. You can see from the picture of the decision tree that the places where the straight line segments (called edges) of the tree end is where the vertices appear in the picture. Each vertex should appear exactly once in the picture. The symbol R stands for the root of the decision tree. Various choices other than R can be used as the symbol for the root.

For any vertex v in a decision tree there is a unique list of edges $(x_1, x_2), (x_2, x_3), ..., (x_k, x_{k+1})$ from the root x_1 to $v = x_{k+1}$. This sequence is called the path to a vertex in the decision tree. The number of edges, k, is the length of this path and is called the height or distance of vertex v to from the root. The path to 22 is $(R, 2), (2, 22)$. The height of 22 is 2.

Because of the decision capability of decision trees, it can be use widely in network fault management (diagnosis)[CZL+04], event correlation and intrusion detection.

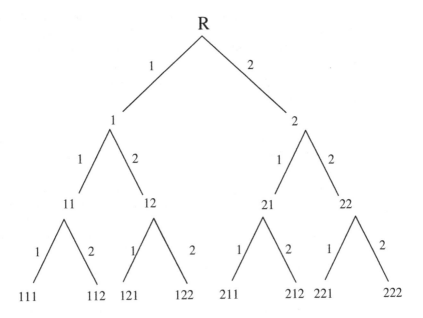

Figure 4.7: Example of a Decision Tree

4.4 Probabilistic Approaches for Network Management

Uncertainty is not only an integral part of human decision-making, but also a fundamental element of the world.

There are various of situations where uncertainty arises since every corner of the world is full of uncertainty. Uncertainty arises from a variety of sources and confuses system designers in a variety of ways. From the engineering view of the world, uncertainty arises when experts concern their own knowledge and when engineers have no exact knowledge of the environment in which their system is supposed to function; uncertainty is inherent in domains engineers normally act in cause their models often abstract out details or unknown facts of the real world that turn out to be relevant to their application later on.

As uncertainty permeates everywhere in the world, any model offering an explicit representation of it would be more accurate and realistic than the one in which uncertainty is disregarded. By modeling uncertainties, certain structural and behavioral aspects of the target system become more visible and understandable, thereby enabling future development steps to be carried out more efficiently and effectively.

Uncertainty comes from the following sources in engineering:

1. Observability: some aspects of the domain are often only partially observable and therefore they must be estimated indirectly through observation. For example, in the management of networks, managers often use their expert knowledge to infer the hidden factors by the obvious evidences when there is no exact or direct management information available by the managed system model.

2. Event correlation: the relations among the main events are often uncertain. In particular, the relationship (dependency) between the observables and nonobservables is often uncertain. Therefore, the nonobservables cannot be inferred from the observables with certainty. The uncertain relation prevents a reasoner from arriving at a unique domain state from observation, which accounts for the guesswork. For example, in a local network the interrupted service in e-mail can be caused by a fault in the Email server. Email client or disconnection in networks, which makes it difficult to determine exactly the location of the root cause when the fault in the e-mail service is reported.

3. Imperfect observations: the observations themselves may be imprecise, ambiguous, vague, noisy, and unreliable. They introduce additional uncertainty to inferring the state of the domain. For complex systems in real life, it is difficult to get full and exact required information; there is always noise and redundancy in the observations.

4. Interpretation: even though many relevant events are observable, very often we only understand them partially. Therefore, the state of the domain must be guessed based on incomplete information. Even though relations are certain in some domains, very often it is impractical to analyze all of them explicitly. Consequently, the state of the domain is estimated from computationally more efficient but less reliable means.

5. Reduction of complexity: another source of uncertainty comes from the most fundamental level of our universe. Heisenberg's uncertainty principle holds that "the more precisely the position is determined, the less precisely the momentum is known in this instant, and vice versa" [Hei27]. Therefore, uncertainty exists within the building blocks of our universe.

In the light of these factors and others, the reasoner's task is not one of deterministic deduction but rather uncertain reasoning. That is, the reasoner must infer the state of the domain based on incomplete and uncertain knowledge (observations) about the domain.

For evolving complex network system, uncertainty is an unavoidable characteristics, which comes from unexpected hardware defects, unavoidable software errors, incomplete management information, and dependency relationship between the managed entities. An effective management system in networks should deal with the uncertainty and suggest probabilistic reasoning for daily management tasks [Din08].

Because of the complexity of managed networks, it is not always possible to build precise models in which it is evident that the occurrence of a given set of alarms indicates a fault on a given element (object) [Din07].

The knowledge of the cause-effect relations among faults and alarms is generally incomplete. In addition, some of the alarms generated by a fault are frequently not made available to the correlation system in due time because of losses or delays in the route from the system element which originates them. Finally, due to the fact that the configuration frequently changes, the more detailed a model is, the faster it will become outdated.

The imprecision of the information supplied by specialists very often causes great difficulties. The expressions "very high," "normal" and "sometimes" are inherently imprecise and may not be directly incorporated to the knowledge basis of a conventional rule-based system.

Fuzzy logic and Bayesian networks are popular models for probabilistic fault management.

4.4.1 Fuzzy Logic

Fuzzy logic is a form of multi-valued logic derived from fuzzy set theory to deal with reasoning that is approximate rather than precise. Just as in fuzzy set theory the set membership values can range (inclusively) between 0 and 1, in fuzzy logic the degree of truth of a statement can range between 0 and 1 and is not constrained to the two truth values true, false as in classic predicate logic. And when linguistic variables are used, these degrees may be managed by specific functions, as discussed below.

Fuzzy logic provides one way to deal with uncertainty and imprecision, which characterize some applications of network management. Fuzzy logic relies on traditional logic, multi-valued logic, probability theory and probabilistic logic as special cases.

The basic concept underlying fuzzy logic is fuzzy sets. Conventional binary logic is based on two possible values: *true* or *false*. When it comes to fuzzy sets, a certain grade of membership, which may assume any value between 0 (when definitely the element does not belong to the set) and 1 (when the element certainly is a member of the set). The concept fuzzy set brings in the novelty that any given proposition does not have to be only *true* or *false*, but that it may be partially true in the interval between 0 and 1.

Although it is possible to empirically attest that a given fuzzy logic system operates according to what is expected, there are still no tools that allow one to prove this [MK94b].

Fuzzy expert systems allow the rules to be directly formulated utilizing linguistic variables such as "very high" or "normal," which rather simplifies the development of the system.

Some applications of fuzzy systems have been implemented in fault management [CH96] [AD99]. [AD99] proposes a fuzzy-logic model for the temporal reasoning phase of the alarm correlation process in network fault management. [CH96] presents a fuzzy expert systems to simplify management within communication networks.

Some researchers argue that all problems that can be solved by means of fuzzy logic may be equally well solved by means of probabilistic models like, for example, Bayesian networks, which have the advantage of relying on a solid mathematical basis.

4.4.2 Bayesian Networks

Bayesian Networks (BNs) are effective means to model probabilistic knowledge by representing cause-and-effect relationships among key entities of a managed system. Bayesian networks can be used to generate useful predictions about future faults and support decisions even in the presence of uncertain or incomplete information. Bayesian networks have been applied in various areas. [Nik00] [WKT+99] [SCH+04] use Bayesian networks in medical diagnosis. [KK94] describes the application of Bayesian networks in the fault diagnosis in Diesel engines. [CSK01] presents methods in distributed Web mining from multiple data streams based on the model of Bayesian networks. [HBR95] describes an application of Bayesian networks in the retrieval of information, according to the users' areas of interest. [BH95] presents a system that utilizes a Bayesian network for debugging very complex computer programs. [Bun96] approaches the utilization of Bayesian networks in coding, representing and discovering knowledge, through some processes that seek new knowledge on a given domain based on inferences on new data or on the knowledge already available [KZ96]. [BDV97] uses Bayesian networks for map learning and [SJB00] applies the model of Bayesian networks to image sensor fusion. [Leu02] [GF01] and [Nej00] use Bayesian networks to model adaptive e-learning environments.

Basic Concepts of Bayesian Networks

The technology with which a system handles uncertain information forms a crucial component of its overall performance. The technologies for modeling uncertainty include Bayesian probability, Dempster-Shafer theory, Fuzzy Logic, and Certainty Factor. Bayesian probability uses probability theory to manage uncertainty by explicitly representing the conditional dependencies between the different knowledge components. It offers a language and calculus for reasoning about the beliefs in the presence of uncertainty. Prior probabilities are thus updated, after new events are observed to produce posterior probabilities. By repeating this process, the implications of multiple sources of evidence can be calculated in a consistent way, and the uncertainties are exploited explicitly to reach an objective conclusion. Bayesian networks provide an intuitive graphical visualization of the knowledge, including the interactions among the various sources of uncertainty.

Bayesian networks also known as Bayesian belief networks, probabilistic networks, or causal networks, are an important knowledge representation techniques in Artificial Intelligence [Pea88] [CDL+99]. Bayesian networks use directed acyclic graphs (DAG) with probability labels to represent probabilistic knowledge.

Bayesian networks can be defined as a triplet (V, L, P), where V is a set of variables (nodes of the DAG) which represent propositional variables of interest, L is the set of causal links among the variables (the directed arcs between nodes of the DAG) which represent informational or causal dependencies among the variables, P is a set of probability distributions defined by: $P = \{p(v \mid \pi(v)) \mid v \in V\}$; $\pi(v)$ denotes the parents of node v. The dependencies are quantified by conditional probabilities for each node given its parents in the network. The network supports the computation of the probabilities of any subset of variables given evidence about another subset.

In Bayesian networks, the information included in one node depends on the information of its predecessor nodes. Any direct predecessor denotes an effect node; the latter represents its causes. Note that an effect node can also act as a causal node of other nodes, where it then plays the role of a cause node.

Causal relations also have a quantitative side, namely their strength. This is expressed by attaching numbers (probabilities) to the links.

Let A be a parent of B. Using probability calculus, it would be natural to let $P(B|A)$ be the strength of the link. However, if C is also a parent of B, then the two conditional probabilities $P(B|A)$ and $P(B|C)$ alone do not give any clue about how the impacts from A and C interact. See Figure 4.8, they may cooperate or counteract in various ways, so we need a specification of $P(B|A, C)$. To each variable B with parents $A_1, \ldots, A_n$, there is attached the potential table $P(B|A_1, \ldots, A_n)$. Note that if A has no parents, then the table reduces to unconditional probabilities $P(A)$.

A Bayesian network provides a complete description and very compact representation of the domain. It encodes Joint Probability Distributions (JPD) in a compact manner. An important advantage of Bayesian networks is the avoidance of building huge JPD tables that include permutations of all the nodes in the network. Rather, for an effect node, only the states of its immediate predecessor need to be examined.

A complete joint probability distribution over n binary-valued attributes requires $2^n - 1$ independent parameters to be specified. In contrast, a Bayesian network over n binary-valued attributes, in which each node has at most k parents, requires at most $2^k n$ independent parameters. To make this concrete, suppose we have 20 nodes ($n = 20$) and each node has at most 5 parent nodes ($k = 5$). Then the Bayesian network requires only 640 numbers, but the full joint probability distribution requires a million. Clearly, such a network can encode only a very small fraction of the possible distributions over these at-

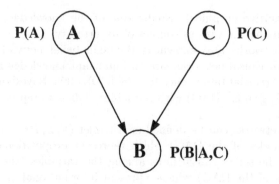

Figure 4.8: Basic Model of Bayesian Networks

tributes, since it has relatively few parameters. The fact that the structure of a BN eliminates the vast majority of distributions from consideration indicates that the network structure itself encodes information about the domain. This information takes the form of the conditional independence relationships that hold between attributes in the network.

To this end, a Bayesian network comprises two parts: a qualitative part and a quantitative part. The qualitative part of a Bayesian network is a graphical representation of the independencies held among the variables in the distribution that is being represented. This part takes the form of an acyclic directed graph. In this graph, each vertex represents a statistical variable that can take one of a finite set of values. Informally speaking, we take an arc $v_i \rightarrow v_j$ in the graph to represent a direct influential or causal relationship between the linked variables v_i and v_j; the direction of the arc $v_i \rightarrow v_j$ designates v_j as the effect or consequence of the cause v_i. Absence of an arc between two vertices means that the corresponding variables do not influence each other directly, and hence, are independent.

Associated with the qualitative part of a Bayesian network is a set of functions representing numerical quantities from the distribution at hand. With each vertex of the graph is associated a probability assessment function, which basically is a set of (conditional) probabilities, describing the influence of the values of the vertex' predecessors on the probabilities of the values of this vertex itself.

Bayesian Networks for Network Management

Because of the dense knowledge representation of Bayesian networks, Bayesian networks can represent large amounts of interconnected and causally linked data as they occur in networks. Generally speaking:

- Bayesian networks can represent knowledge in depth by modeling the functionality of the transmission network in terms of cause-and-effect relationship among network components and network services.

- Bayesian networks are easy to extend in design because they are graph based models. Hence Bayesian networks are appropriate to model the problem domain in networks, particularly in the domain of fault propagation.

- Bayesian networks come with a very compact representation. A complete Joint Probability Distribution (JPD) over n binary-valued attributes requires $2^n - 1$ independent parameters to be specified. In contrast, a Bayesian network over n binary-valued attributes, in which each node has at most k parents, requires at most $2^k n$ independent parameters. It is clear that such a network can encode only a very small fraction of the possible distributions over these attributes, since it has relatively few parameters. The fact that the structure of a Bayesian network eliminates the vast majority of distributions and indicates that the network structure itself encodes information about the domain. This information takes the form of the conditional independence relationships that hold between attributes in the network.

- Bayesian networks have the capability of handling noisy, transient, and ambiguous data, which is unavoidable in complex networks, due to their grounding in probability theory.

- Bayesian networks have the capacity to carry out inference on the presence of a network from the combination of:

  - statistical data empirically surveyed during the network functioning;
  - subjective probabilities supplied by specialists and
  - information (that is, "evidences" or "alarms") received from the distributed systems.

- Bayesian networks provide a compact and well-defined problem space because they use an exact solution method for any combination of evidence or set of faults. Through the evaluation of a Bayesian network, it is possible to obtain approximated answers, even when the existing information is incomplete or imprecise; as new information becomes available, the Bayesian networks allow a corresponding improvement in the precision of the correlation results.

- Bayesian networks are abstract mathematic models. In network management, Bayesian networks can be designed on different levels on different management intention or based on particular application or services. For example, when a connection service is considered, the physical topology is the basis for the construction of a Bayesian network. While a distributed service is taken into account, the logic dependency based on certain service between managed objects will act as the foundation in construction of a Bayesian network.

To give an example, consider a client-server communication problem where client stations set up connections with a server station via an Ethernet. Suppose a user complains that his client station cannot access the server. Possible causes of this failed access (AF) might be due to a server fault (SF), network congestion (NC), or link fault (LF). Network congestion is usually caused by high network load (HL) and evidenced by low network throughput and long packet delays and even losses. We also assume that a link fault (LF) will sometimes trigger the link alarm (LA). This example is illustrated in Figure 4.9.

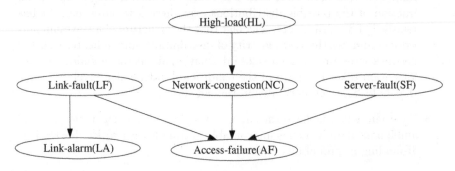

Figure 4.9: An Example of a Bayesian Network in Network Management

Assume for simplicity that all the random variables are binary valued. For a random variable, say, HL, we use hl to denote that the value is true and $\overline{hl}$ to denote that the value is false. If the systems manager receives a user complaint that the access has failed (af), and the link alarm (la) was issued, she can calculate the conditional probabilities of link-fault $p(lf|af, la)$, network congestion $p(nc|af, la)$ and server fault $p(sf|af, la)$, respectively.

We can see that the basic computation in Bayesian networks is the computation of every node's belief given all the evidence that has been observed so far. In general, such computation is NP-Hard [Coo90].

Some applications of Bayesian networks for fault management have been reported. [DLW93] provides a probabilistic approach to fault diagnosis in linear lightwave networks based on Bayesian networks model. [WS93b] uses Bayesian networks in diagnosing connectivity problems in communication systems. [HJ97a] utilizes Bayesian network models to accomplish early detection by recognizing deviation from normal behavior in each of the measurement variables in different network layers. [DKB+05] presents Bayesian networks based technique for backward inference in fault management, this approach can efficiently trace the root causes of faults in complex networks.

4.5 Web-Based Network Management

The fast growth of the WWW (World Wide Web) made it possible to develop a wide range of web-based network management architectures.

4.5.1 Web-Based Network Management

Basically there are three Network Management System Frameworks: centralized, hierarchical, and distributed.

- In the centralized framework, a unique host keeps an eye on all IT systems. Although simple, it shows problems such as a high concentration of fault probability on a single element and low scalability.

- In the hierarchical framework, each domain manager has to monitor its own domain only. The highest-level domain manager make periodical polls to the lowest-level domain managers.

- In distributed framework, like peer-to-peer, multiples managers, each manages its own domain, communicates with the others on a peer system. A manager from any domain can poll information to another manager from another domain.

Distributed framework is appropriate approach in managing enlarging networks. Web-based network management is the ability to monitor and actively manage a network, regardless of the location of the network or the network manager, in real time, using the Internet as an access vehicle.

Web-based network management (WebNM) uses the World Wide Web technologies to manage network systems with the benefits of platform-independence, uniform management interface, and reduced costs. The most important advantage is that it allows network managers administrate the network at any network management check-point. WebNM abstracts the traditional SNMP network management framework from the network managers by using HTTP as the management information transfer protocol among network managers and managed objects. This provides network management solutions to those enterprise networks with multi-vendor network devices and heterogeneous network environment. With this system, network managers are able to manage the entire network remotely through the web browser.

More technically, it is a system that allows the Management Information Systems (MIS) department to leverage Web technology to extend the function of its network management system. It lets a user extend the existing monitoring, configuration, and troubleshooting functions of the corporate network administrators by taking advantage of the natural efficiencies of the browser.

Browsers are that family of handy tools that allow users to move from one World Wide Web (WWW) site to another with ease and efficiency. Browsers read Hypertext Markup Language (HTML) and then link the user's computer with the desired site. The WWW site can be anything from a corporate home page to a device on the corporate network.

There are some advantages of using web-based network management:

- Ease of access. Web-based network management allows for remote management capabilities. Access to Network Management Information without the use of any special Network Manager. The dream of "anywhere, anytime" management has been realized. The work comes to the worker, rather than the worker having to move to the work.

- Platform independency. Web-based network management system provides a simple unified web-based user interface. The management framework and its location are independent. There is no need to scatter Sun workstations or X-terminals in locations that someone considers to be strategic sites today, only to find that tomorrow they are not as convenient or ubiquitous as was once thought. Web-based network management enables infinite vendor support, can facilitate the opening up of the network for plug-and-play, can easily integrate with operators other business systems. Technically, new management features can be rapidly developed and installed, and eliminate costly software upgrades.

- Ease of use. Web-based network management system enables minimal training of staff. It provides intuitive, graphical interface for management operation. User interaction is via pages defined by the developer, who can use an easy mechanism to define task-oriented pages. Simply by adding a bookmark to a computer browser, a network manager can hop to the management component of the system under consideration. Network managers can interact with each other easily for cooperation tasks by available web techniques. For access control, it maintains a list of authorized users, who are allowed to specify the accessibility. Technically, web techniques enables accessing group related MIB field variables by one entity. Meanwhile, using web-based technologies for enterprise management is where the industry is going.

Figure 4.10 illustrates the architecture of an example web-based network management system. Web-based network management is widely used in heterogeneity network configuration management.

4.5.2   Web-Based Enterprise Management

Web-Based Enterprise Management (WBEM) is a set of systems management technologies developed to unify the management of distributed computing environments. WBEM is based on Internet standards and Distributed Management Task Force (DMTF) open standards: Common Information Model (CIM) infrastructure and schema, CIM-XML, CIM operations over HTTP, and WS-Management. Other systems management approaches are remote shells, proprietary solutions, and network management architectures

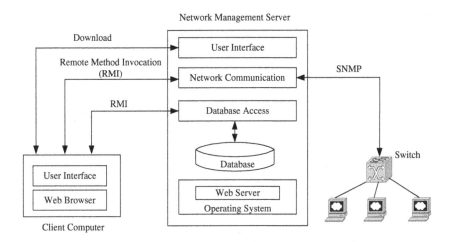

Figure 4.10: Web-Based Network Management

like SNMP.

Key features of WBEM technology include:

- remote management of applications

- management of several instances of an application as a single unit

- standard interface for remote application management across different applications

- decoupling of application management from the client

- "publishing" of key information about an application to other applications

Web-based Enterprise Management Architecture

To understand the WBEM architecture, consider the components that lie between the operator trying to manage a device (configure it, turn it off and on, collect alarms, etc.) and the actual hardware and software of the device: the operator will presumably be presented with some form of graphical user interface (GUI), browser user interface (BUI), or command line interface (CLI). The WBEM standard really has nothing to say about this interface (although a CLI for specific applications is being defined): in fact it is one of the strengths of WBEM that it is independent of the human interface since human interfaces can be changed without the rest of the system needing to be aware of the changes. The GUI, BUI, or CLI will interface with a WBEM client through a small set of application programming interfaces (API). This client will find the WBEM server for the device being managed (typically on the device itself) and construct an XML message with the request.

The client will use the HTTP (or HTTPS) protocol to pass the request, encoding in CIM-XML, to the WBEM server the WBEM server will decode the incoming request, perform the necessary authentication and authorization checks, and then consult the previously created model of the device being managed to see how the request should be handled. This model is what makes the architecture so powerful: it represents the pivot point of the transaction with the client simply interacting with the model and the model interacting with the real hardware or software. The model is written using the Common Information Model standard and the DMTF has published many models for commonly managed devices and services: IP routers, storage servers, desktop computers, etc., for most operations, the WBEM server determines from the model that it needs to communicate with the actual hardware or software. This is handled by so-called providers: small pieces of code which interface between the WBEM server (using a standardized interface known as CMPI) and the real hardware or software. Because the interface is welldefined and the number of types of call is small, it is normally easy to write providers. In particular, the provider writer knows nothing of the GUI, BUI, or CLI being used by the operator.

WBEM products interoperate in a multisystem environment by using technologies designed around WBEM standards. When Microsoft developed its common, object-based information model called HMMS (HyperMedia Management Schema), it was adopted by the DMTF and evolved into the Common Information Model (CIM), now published as the CIMv2 schema. The DMTF plans to develop and publish additional WBEM standards so that WBEM products that manage heterogeneous systems, regardless of their instrumentation mechanisms, can be developed by management vendors.

Figure 4.11 presents the architecture of web-based enterprise management.

4.6 Agent Techniques for Network Management

4.6.1 Introduction of Agent

The agent concept, widely proposed and adopted within both the telecommunications and Internet communities, is a key tool in the creation of an open, heterogeneous, and programmable network environment. Agents enhance the autonomy, intelligence, and mobility of software objects and allow them to perform collective and distributed tasks across the network. The term "agent" is used in many different contexts. Thus, there is no single (software) agent definition that exists today. However, agents can be discussed by means

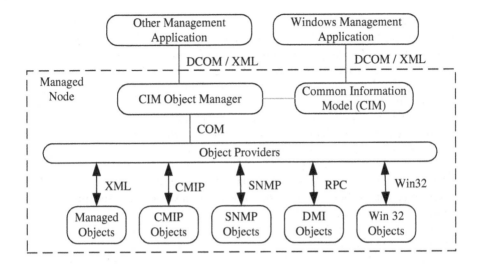

Figure 4.11: Web-Based Enterprise Management

of several attributes to distinguish them from ordinary code. It is not essential that all of these attributes be present in a given agent implementation. The single attribute that is commonly agreed upon is autonomy. Hence, an agent can be described as a software component that performs a specific task autonomously on behalf of a person or organization. Alternatively, an agent can be regarded as an assistant or helper that performs routine and complicated tasks on a user's behalf. In the context of distributed computing, an agent is an autonomous software component that acts asynchronously on the user's behalf. Agent types can be broadly categorized as static or mobile.

Based on these definitions, an agent contains some amount of artificial intelligence, ranging from predefined and fixed rules to self-learning mechanisms. Thus, agents may communicate with the user, system resources, and other agents to complete their task. The agents may move from one system to another to execute a particular task.

Agent Techniques

The agent technologies broadly consist of an agent communication language, a distributed processing environment (DPE), and a language for developing agent-based applications. To cooperate and coordinate effectively in any multiagent system, agents are required to communicate with each other. A standard communication language is essential if agents are to communicate with each other. There are two main categories of agent communication: basic communication and communication based on standardized agent communication languages (ACLs). Basic communication is restricted to a finite set of signals. In a multiagent system, these signals have the capability of invoking other agents for a desired action, whereas communication based on ACLs involves the use of a software language for communication between agents. For example, Foundation for Intelligent Physical Agents (FIPA) ACL is mainly based on the Knowledge Query and Manipulation Language (KQML). KQML can be considered to consist of three layers: the content, message, and communication layers. The content layer specifies actual proportions of the message. At the second level, KQML provides a set of performatives that constitutes the message layer (e.g., ask, tell, reply, etc.). A KQML message could contain Prolog code, C code, natural language expressions, or whatever, as long as it can be represented in ASCII. This layer enables agents to share their intentions with others. The communications layer defines the protocol for delivering the message and its encapsulated content.

Languages such as Smalltalk, Java, C++, or C can be used for the construction of agent-based applications. Choice of the language depends on the type of agent functionality. For example, Java or Telescript is suitable for implementing mobile agents, whereas tool command language (TCL) is more appropriate for interface agents.

To provide a standard agent operational environment in relation to underlying communications infrastructure, creation of a DPE is necessary. The creation of a DPE provides physical communication transparency to the agents via Java Remote Method Invocation (RMI) or Common Object Request Broker Architecture (CORBA). These provide the supporting infrastructure for a DPE where agents can operate. It may involve the creation of a platform-independent agent environment generated by the execution of, e.g., Java Virtual Machines at hardware nodes. The majority of current communication system architectures employ the client/server method that requires multiple transactions before a given task can be accomplished. This can increase signaling traffic throughout the network. This problem can rapidly escalate in an open network environment that spans multiple domains. As an alternative solution, mobile agents can migrate the computations or interactions to the remote host by moving the execution there.

Decentralized network management systems are a possible answer to scalability problems presented by the traditional server-centric systems. They introduce advantages in scalability, interoperability, survivability and concurrent diagnosis.

Distributed management architectures are an alternative for the manager-centric scenario imposed by architectures such as SNMP and CMIP, providing better scalability and resource utilization. However, the implementation of this distributed architecture is a complex engineering task, incurring in high level of development costs and risks. As a result of using agent-based development for network management, the guidelines for the application become:

- high degree of adaptability, which is inherent to the agent technology, being one agent an environment aware and responsive piece of software;

- code mobility, as the self-contained agents represent a simple abstraction for software move between element;

- module reusability, as each agent can implement a module function and multiple agents interact during problem resolution, and;

- self-generation, due to the agents self-contained features it is easier to implement agents that create new agents customized for specific jobs.

There are two major research areas in DAI (Distributed Artificial Intelligence) for management systems: the first deals with the use of mobile code or, more specifically, the mobile agents; the second covers the design of multiagent (intelligent agent) systems in largely distributed architectures.

4.6.2 Mobile Agents

Mobile Agent, namely, is a type of software agent, with the feature of autonomy, social ability, learning, and most importantly, mobility.

When the term "mobile agent" is used, it refers to a process that can transport its state from one environment to another, with its data intact, and be capable of performing appropriately in the new environment. Mobile agents decide when and where to move. Movement is often evolved from RPC methods. A mobile agent accomplishes a move through data duplication. When a mobile agent decides to move, it saves its own state and transports this saved state to the new host and resumes execution from the saved state.

In contrast to the Remote evaluation and Code on demand paradigms, mobile agents are active in that they can choose to migrate between computers at any time during their execution. This makes them a powerful tool for implementing distributed applications in a computer network.

Some advantages which mobile agents have over conventional agents:

- Move computation to data, reducing network load

- Asynchronous execution on multiple heterogeneous network hosts

- Dynamic adaptation – actions are dependent on the state of the host environment

- Tolerant to network faults – able to operate without an active connection between client and server

- Flexible maintenance – to change an agent's actions, only the source (rather than the computation hosts) must be updated

The common applications of mobile agents include:

- Resource availability, discovery, monitoring

- Information retrieval can be used in fetching the system information in client/server paradigm

- Network management – by using mobile agents we can monitor the throughput of remote machine in terms of network parameters

- Copying files in a server-client paradigm or backing up of data on remote machines

- Dynamic software deployment can increase portability, making system requirements less influential.

Mobility agent could solve the problems that the centralize systems are facing, such as the restriction of upgrading and the inflexibility of the static structure. After all, mobility agent still has some problems when handling with the dynamic change of the traffic and the topology features of the modern network.

Figure 4.12 illustrates the mobile agent-based distributed network management framework [DLC03].

One of the great advantages of mobile agent paradigms in network management lies in the aspect of decentralization of the manager figure. The agents are conveniently distributed in the environment and execute tasks that would normally be managers, responsibility. Mobile agents can decentralize processing and control thus improving management efficiency. Some advantages that justify mobile agents utilization in network management are [KW+04]:

- Cost reduction: inasmuch as management functions require the transfer on the network of a great volume of data, it may be better to send an agent to perform the task directly on network elements where data are stored.

- Asynchronous processing: an agent can be sent through the network, performing its tasks on other nodes. While the agent is out of its home node, this node can be out of operation.

- Distributed processing: low capacity computers can be efficiently used in order to perform simple management tasks, distributing processing previously concentrated on the management station.

- Flexibility: a new behavior for the management agents can be determined by the management station, which sends a mobile agent with a new execution code, substituting the old one in realtime.

Comparing to classical management, the mobile agents are more than a simple transport mechanisms for collecting data. Mobile agents can carry out any computation, theoretically meeting any need, superimposing the static characteristics of the client-server models of network management. Of course, this very feature imposes security problems on the architecture while one

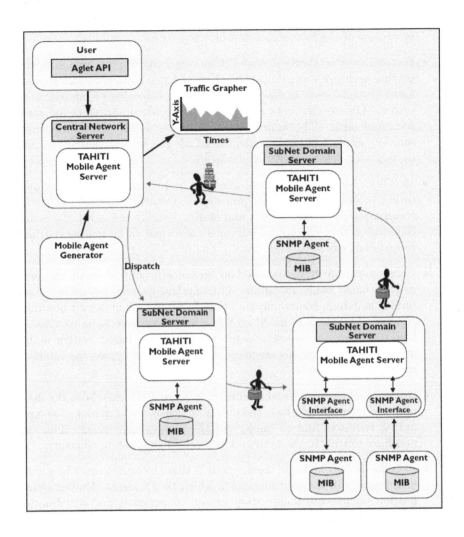

Figure 4.12:   A Framework of a Mobile Agent-Based Distributed Network Management System

agent can potentially carry out harmful or illicit activities while running on the network element.

For successful implementation of the mobile agent systems for network management, there are some requirements that should be met:

- Security: The security of mobile agents is concerned with the requirements and actions of protecting the AEE (Agent Execution Environment) from malicious agents, protecting agents from a malicious AEE, and protecting one agent from another. It is also concerned with the protection of the communication between AEEs.

- Portability: Portability deals with heterogeneity of platforms and with porting agent code and data to work on multiple platforms. Before Java, agent systems such as Agent Tcl and ARA depended on an operating system (OS)-specific core and language-specific interpreters for agent code and data. The Java Virtual Machine (JVM) presents a better solution, as platform-independent bytecode can be compiled just in time to deal with heterogeneity of platforms.

- Mobility: Most systems mentioned in the previous section use application protocols on top of Transmission Control Protocol (TCP) for transportation of agent codes and states. Usually the agent states and codes are converted into an intermediate format to be transported and restarted at the other end.

- Communication: Systems based on Java mostly support event, message, or RMI-based communication. This method has the advantage of interface matching communication that facilitates client/server bindings. However, because it is based on RPC, interfaces have to be established before a proper connection is established. Java based system makes use of the homogeneous language environment to bypass the interface incompatibility issue.

- Resource management and discovery: For agent systems, both the manager and agent should have the ability to discover and manage the computing resources and to implement the management tasks. This can also have impact to the system performance and system efficiency.

- Identification: The basic requirement is that the agents must be identified uniquely in the environment in which they operate. Proper identification permits communication control, cooperation, and coordination of agents to take place.

- Control: Control in agent systems is based on the coordination of mobile agent activities. Furthermore, it provides means through which agents may be created, started, stopped, duplicated, or instructed to self-terminate.

- Data management: In agent-based systems, mobile agent is embedded with status and data, agent manager holds sufficient data to support the running environment and service operation. All these system data and running data should be managed in order, in reliability, and in security.

4.6.3 Intelligent Agents

In artificial intelligence, an intelligent agent (IA) is an entity that observes and acts upon an environment (i.e., it is an agent) and directs its activity toward achieving goals (i.e., it is rational). Intelligent agents may also learn or use knowledge to achieve their goals. They may be very simple or very complex: a reflex machine such as a thermostat is an intelligent agent, as is a human being, as is a community of human beings working together toward a goal.

Intelligent agents are often described schematically as an abstract functional system similar to a computer program. For this reason, intelligent agents are sometimes called abstract intelligent agents (AIA) to distinguish them from their real-world implementations as computer systems, biological systems, or organizations. Some definitions of intelligent agents emphasize their autonomy and so prefer the term autonomous intelligent agents. Still others considered goal-directed behavior as the essence of rationality and so preferred the term rational agent.

Intelligent agents have been defined many different ways. However, IA systems should exhibit the following characteristics:

- accommodate new problem solving rules incrementally

- adapt online and in real time

- be able to analyze itself in terms of behavior, error, and success

- learn and improve through interaction with the environment (embodiment) learn quickly from large amounts of data

- have memory-based exemplar storage and retrieval capacities

- have parameters to represent short- and long-term memory, age, forgetting, etc.

The program intelligent agent maps every possible percept to an action. It is possible to group agents into five classes based on their degree of perceived intelligence and capability:

- Simple reflex agents: Simple reflex agents acts only on the basis of the current percept. The agent function is based on the condition-action rule: if condition then action rule. This agent function only succeeds when the environment is fully observable. Some reflex agents can also contain information on their current state that allows them to disregard conditions whose actuators are already triggered.

- Model-based agents: Model-based agents can handle partially observable environments. Its current state is stored inside the agent maintaining some kind of structure that describes the part of the world that cannot be seen. This behavior requires information on how the world behaves and works. This additional information completes the "World View" model.

- Goal-based agents: Goal-based agents are model-based agents that store information regarding situations that are desirable. This allows the agent a way to choose among multiple possibilities, selecting the one that reaches a goal state.

- Utility-based agents: Goal-based agents only distinguish between goal states and nongoal states. It is possible to define a measure of how desirable a particular state is. This measure can be obtained through the use of a utility function that maps a state to a measure of the utility of the state.

- Learning agents: Learning has an advantage that it allows the agents to initially operate in unknown environments and to become more competent than its initial knowledge alone might allow.

Unlike mobile agent, an intelligent agent does not need to be given task instructions to function, rather just high-level objectives. The use of intelligent agents completely negates the need for dedicated manager entities, as intelligent agents can perform the network management tasks in a distributed and coordinated fashion, via inter agent communications.

Figure 4.13 illustrates the framework of intelligent agent.

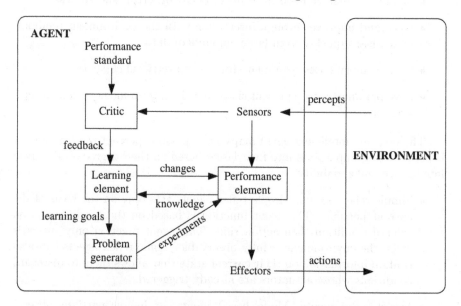

Figure 4.13: Framework of Intelligent Agents

Many researchers believe intelligent agents are the future of network management, since there are some significant advantages in using intelligent agents for network management.

- Firstly, intelligent agents would provide a fully scalable solution to most areas of network management. Hierarchies of intelligent agents could each assume a small task in their local environment and coordinate their efforts globally to achieve some common goal, such as maximizing the overall network utilization.

- Secondly, data processing and decision making are completely distributed, which alleviates management bottlenecks as seen in centralized network management solutions. In addition, the resulting management system is more robust and fault tolerant, as the malfunction of a small number of agents would have no significant impact on the overall management function.

- Thirdly, the entire management system is autonomous; network administrators would only need to provide service-level directives to the system.

- Lastly, intelligent agents are self-configuring, self-managing, and self-motivating. It is ultimately possible to construct a management system that is completely self-governed and self-maintained. Such a system would largely ease the burden of network management routines that a systems administrator has to currently struggle with.

The application of intelligent agents to network management is still in its infancy, and many difficult issues still remain unsolved. As applications utilizing intelligent agents arise in network management, the problem of managing these intelligent agents will also become increasingly important. These self-governing agents cannot simply be allowed to roam around the network freely and access vital resources. Currently, it is still very difficult to design and develop intelligent agent platforms. This is mostly because very little real-life practice with intelligent agents exists today. As intelligent agents are empowered with more intelligence and capabilities, their size will become an increasing concern for network transport. Furthermore, agent-to-agent communication typically uses the Knowledge Query Manipulation Language (KQML). KQML wastes a substantial amount of network resources, as its messages are very bulky. Lastly, similar to mobile agents, the security of intelligent agents is a big barrier in their applications. In particular, the following questions remain unanswered: Who takes care of agent authentication? Can agents protect themselves against security attacks? Can agents keep their knowledge secret? How much access rights should agents have over network resources?

4.7   Distributed Object Computing for Network Management

4.7.1   Introduction of Distributed Object Computing

Distributed objects are software modules that are designed to work together but reside either in multiple computers connected via a network or in different processes inside the same computer. One object sends a message to another object in a remote machine or process to perform some task. The results are sent back to the calling object.

Middleware is a class of software technologies designed to help manage the complexity and heterogeneity inherent in networks. It has two major components: OMG (Object Management Group) and CORBA (the Common Object Request Broker Architecture). The OMG describes how objects distributed across a heterogeneous environment can be classified. It also distinguishes interactions between those objects. On the other hand, CORBA provides communication and activation infrastructure for these distributed objects by using IDL and inter-ORB protocols.

4.7.2   Distributed Object Computing for Network Management

Network applications communicate with one another to deliver highlevel services, such as controlling a multimedia communication session or managing a set of physical or data resources. The creation of QoS-based communication services and the management of network resources can be conflicting objectives that network designers have to meet. Middleware technology provides a logical communication mechanism among network entities by hiding the underlying implementation details of different hardware and software platforms. There are different types of middleware, e.g., distributed tuples, remote distributed object middleware (CORBA, Distributed Component Object Model, or DCOM), etc.

CORBA from the OMG has been widely considered to be the choice architecture for the next generation of network management. OMG is based on its Object Management Architecture (OMA), and CORBA is one of its key components. The most significant feature of CORBA is its distributed object-oriented (OO) architecture. This extends the OO methodology into the distributed computation environment. Its advantages also include interoperability and independence of platform. The Object Request Broker (ORB) is used as the common communication mechanism for all objects. OMG defines a stub as the client-end proxy and a skeleton as the server-end proxy. Using the ORB, a client can transparently invoke a method on a server object, which can be on the same machine or across the network. From the CORBA 2.0 release onward, every ORB is expected to understand at least the Internet Inter-ORB Protocol (IIOP), which is layered on top of TCP.

Until now, most of the research has been performed using CORBA to integrate incompatible legacy network management systems (e.g., SNMP and CMIP) and the definitions of specific CORBA services (e.g., notification, log). In addition to the basic features of CORBA, the OMG has defined some services that can provide a more useful API. Among them, the naming and life cycle services are basic CORBA services that are used by almost all CORBA applications. There are already some free and commercial versions of CORBA-SNMP and CORBA-CMIP gateways available.

Distributed object computing (DOC) uses an object-oriented (OO) methodology to construct distributed applications. Its adaptation to network management is aimed at providing support for a network management architecture, integration with existing solutions for heterogeneous network management, and providing development tools for network management components.

Distributed object computing enables a distribution of services and applications in a seamless and location transparent way by separating object distribution complexity from network management functionality concerns. Another advantage of this separation of concerns is the ability to provide multiple management communication protocols accessed via a generalized abstract programming interface (API), fostering interoperability of heterogeneous network management protocols, such as SNMP for IP networks and Common Management Information Protocol (CMIP) for telecommunication networks. In addition, DOC provides a distributed development platform for rapid implementation of robust, unified, and reusable services and applications.

Distributed object computing, in general, and the Common Object Request Broker Architecture (CORBA), in particular, are well-received technologies for developing integrated network management architectures with object distribution. The success of CORBA as an enabling distributed object technology can be attributed to the fact that CORBA specifies a well-established support environment for efficient run-time object distribution and a set of support services. On this basis DOC is useful as an integration tool for heterogeneous network management domains, and for extending deployed network management architectures. However, DOC still uses static object distribution. It does not have the flexibility that code mobility offers. Furthermore, DOC requires dedicated and heavy run-time support, which may not always be feasible on every device in a network. These issues restrict its range of application.

The Telecommunications Management Network (TMN) has been developed as the framework to support administrations in managing telecommunications networks and services. It suggests the use of OSI Systems Management (OSI-SM) as the key technology for management information exchanges. The latter follows an object-oriented approach in terms of information specification but leaves aspects related to the software structure of relevant applications unspecified. Distributed object technologies, such as CORBA, address the use of software Application Program Interfaces (APIs) in addition to interoperable protocols. Their ease of use, generality, and ubiquity implies that they might be also used in telecommunications management systems [Pav00].

4.8 Active Network Technology for Network Management

4.8.1 Introduction of Active Network

Active networking is a communication pattern that allows packets flowing through a telecommunications network to dynamically modify the operation of the network.

The active network architecture is composed of execution environments (similar to a Unix shell that can execute active packets), a node operating system capable of supporting one or more execution environments. It also consists of active hardware, capable of routing or switching as well as executing code within active packets. This differs from the traditional network architecture, which seeks robustness and stability by attempting to remove complexity and the ability to change its fundamental operation from underlying network components. Network processors are one means of implementing active networking concepts. Active networks have also been implemented as overlay networks.

Active networking allows the possibility of highly tailored and rapid "real-time" changes to the underlying network operation. This enables such ideas as sending code along with packets of information allowing the data to change its form (code) to match the channel characteristics. The use of real-time genetic algorithms within the network to compose network services is also enabled by active networking.

Active networks enable network elements, primarily routers and switches, to be programmable. Programs that are injected into the network are executed by the network elements to achieve higher flexibility for networking functions, such as routing, and to present new capabilities for higher layer functions by allowing data fusion in the network layer. In active networks, routers and switches run customized services that are uploaded dynamically from remote code servers or from active packets. The characteristic of activeness is threefold. From a device view, the device's services and operations can be dynamically updated and extended actively at run time. From a network provider view, the entire network resources can be provisioned and customized actively on a per customer basis. In the view of the users of networks, the allocated resources can be configured actively based on user application needs [BK02].

Active networks combined with code mobility present an effective enabling technology for distributing management tasks to the device level. Not only management tasks can be off-loaded to individual network devices, but also the suppliers of management tasks need no longer be manager entities. Such a solution provides full customization, device-wise, service provider-wise, and user-wise; it provides the means for a distributed process across all network devices; it is interoperable across platforms via device-independent active code; it fosters user innovation and user-based service customization; it accelerates new service and network technology deployment, bypassing standardization process and vendor consensus; it allows for fine grained resource allocation based on individual service characteristics.

4.8.2 Active Network Management

The programmability introduced by the active networks may represent the leap toward a pervasive network management. The use of network services is widespread in today's society. The importance of services requires that they are guaranteed by means of a continuous monitoring for immediate interventions in case of fault. The concept of Resilience in Communications Networks has become an indispensable feature of the network architecture design. Resilience is commonly defined as the set of mechanisms that are able to cope with network failures. In the field of network management, this is well known as proactive management. Proactive management operates on the ground of symptoms that predict negative events in order to avoid them.

In this scenario active networks may be a suitable instrument for the implementation of a complete management system. The actual benefits that management applications may obtain from the active networks adoption fall into the following categories [BCF+02]:

- availability of information held by intermediate nodes;

- data processing capability along the path;

- adoption of distributed and autonomous strategies into the network.

The above features completely answer the network management requirements. Mobile agents can be encapsulated and transported in the active code of application capsules. They can retrieve and extract pieces of information held by intermediate nodes in a more effective way than through remote queries from the application itself. For instance, an agent could make use of an active code to look up the MIB objects of an intermediate node and select some entries according to a given criterion. It can either send such extracted information back to the application, or it can use the information to take timely decisions autonomously from the application. More examples can be found in other network management issues, such as congestion control, error management, or traffic monitoring. A meaningful example is the customization of the routing function. A mobile agent could be devoted to the evaluation of the path for the application's data flow, according to the user's QoS specification. Each application could set up its own control policy or exploit a common service (the default per-hop forward function). Active networks applications can easily implement distributed strategies by spreading management mobile agents in the network. The introduction of network node programmability makes the network system one single knowledge base, which is also capable of producing new information. This happens, for instance, when new actions are generated deductively from the resolution of previously stored data with occurrences of particular events, thus allowing the inference of new events and the triggering of codified measures.

Figure 4.14 presents the Active Network Framework [BK02] .

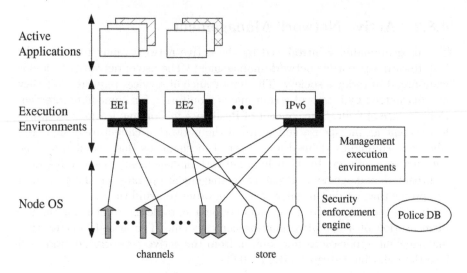

Figure 4.14: Active Network Framework

Active network framework includes following components:

- Active Application (AA): The active network application.

- Execution Environment (EE): Analogous to a Unix shell in which to execute a packet.

- Node Operating System (Node OS): Operating System support for Execution Environments.

The logical flow of packets through an active node is shown in Figure 4.15 [BK02].

Active networks represent a "quantum step" by providing a programmable interface in network nodes, they expose the resources, mechanisms, and policies underlying this increased functionality and provide mechanisms for constructing or refining new services from those elements. In short, active networks support dynamic modification of the network's behavior as seen by the user. Such dynamic control is potentially useful on multiple levels:

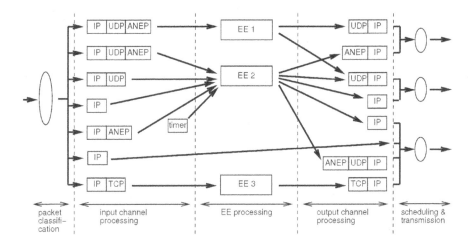

Figure 4.15: Packet Flow through an Active Node

- From the point of view of a network service provider, active networks have the potential to reduce the time required to develop and deploy new network services. The shared infrastructure of the network presently evolves at a much slower rate than other computing technology. One consequence of being able to change the behavior of network nodes on the fly is that service providers would be able to deploy new services quickly, without going through a lengthy standardization process.

- At a finer level of granularity, active networks might enable users or third parties to create and tailor services to their particular applications and even to current network conditions. Although it seems likely that most end users would not write programs for the network, after all, individual users can, in principle, program their PCs, but how many do? It is easy to imagine individuals customizing services by choosing options in code provided by third parties. Indeed, this prospect should appeal to network providers as well, because it enables them to charge more for such value-added services.

- Networks are expensive to deploy and administer. For researchers, a dynamically programmable network offers a platform for experimenting with new network services and features on a realistic scale without disrupting regular network service.

In the literature, there are two general approaches for realizing active networks: the programmable switch approach and the capsule approach. The programmable switch approach uses an out-of-band channel for code distribution. The transportation of active code is completely separated from regular data traffic. This approach is easier to manage and secure, as the active code is distributed via private and secure channels. It is suitable for network administrators configuring network components. On the other hand, the capsule approach packages active code into regular data packets. The active code is sent to active nodes via regular data channels. This approach allows open customization of user-specified services. The downside of this approach is, however, that it is prone to security threats. [BP02] analyzed the benefits of active networks to enterprise network management.

Some recent work has been done on exploring active networks for network management, such as the Virtual Active Network (VAN) proposal [Bru00] and the agent-based active network architecture [HC02]. However, security remains a major roadblock for practical application of active networks. Not only the integrity of network resources and user data has to be kept, but also the content of user data must remain confidential.

Besides security, resource provisioning and fault tolerance are the other two major issues that need to be addressed in active networks.

4.9 Bio-inspired Approaches

4.9.1 Bio-inspired Computing

Bio-inspired computing (biologically inspired computing) is a field of study that loosely knits together subfields related to the topics of connectionism, social behavior and emergence. It is often closely related to the field of artificial intelligence, as many of its pursuits can be linked to machine learning. It relies heavily on the fields of biology, computer science, and mathematics. Briefly put, it is the use of computers to model nature and simultaneously the study of nature to improve the usage of computers. Biologically inspired computing is a major subset of natural computation.

The way in which bio-inspired computing differs from traditional artificial intelligence (AI) is in how it takes a more evolutionary approach to learning, as opposed to the what could be described as "creationist" methods used in traditional AI. In traditional AI, intelligence is often programmed from above: the programmer is the creator and makes something and imbues it with its intelligence. Bio-inspired computing, on the other hand, takes a more bottom-up, decentralized approach; bio-inspired techniques often involve the method of specifying a set of simple rules, a set of simple organisms that adhere to those rules, and a method of iteratively applying those rules. After several generations of rule application, it is usually the case that some forms of complex behavior arise. Complexity gets built upon complexity until the end result is something markedly complex and quite often completely counterintuitive from what the original rules would be expected to produce.

Natural evolution is a good analogy to this method. The rules of evolution (selection, recombination/reproduction, mutation and more recently transposition) are in principle simple rules, yet over thousands of years have produced remarkably complex organisms. A similar technique is used in genetic algorithms.

Swarm Intelligence

Swarm intelligence (SI) is a type of artificial intelligence based on the collective behavior of decentralized, self-organized systems. The expression was introduced in the context of cellular robotic systems.

SI systems are typically made up of a population of simple agents interacting locally with one another and with their environment. The agents follow very simple rules, and although there is no centralized control structure dictating how individual agents should behave, local, and to a certain degree random, interactions between such agents lead to the emergence of "intelligent" global behavior, unknown to the individual agents. Natural examples of SI include ant colonies, bird flocking, animal herding, bacterial growth, and fish schooling.

There are five fundamental principles of swarms:

- Proximity: a swarm can make simple calculations about time and space.

- Quality: a swarm can react to indications by the environment.

- Diverse response: activities can be performed in different ways.

- Stability: not every environmental change modifies the swarm.

- Adaptability: a swarm can change its behavior if that seems promising.

These properties have been implemented in many agent-based models.

The application of swarm principles to robots is called swarm robotics, while "swarm intelligence" refers to the more general set of algorithms. "Swarm prediction" has been used in the context of forecasting problems.

Example algorithms in swarm intelligence:

- Ant colony optimization

  Ant colony optimization (ACO) is a class of optimization algorithms modeled on the actions of an ant colony. Artificial "ants" (simulation agents) locate optimal solutions by moving through a parameter space representing all possible solutions. Real ants lay down pheromones directing each other to resources while exploring their environment. The simulated "ants" similarly record their positions and the quality of their solutions, so that in later simulation iterations more ants locate better solutions. One variation on this approach is the bees algorithm, which is more analogous to the foraging patterns of the honeybee.

- Particle swarm optimization

  Particle swarm optimization (PSO) is a global optimization algorithm for dealing with problems in which a best solution can be represented as a point or surface in an n-dimensional space. Hypotheses are plotted in this space and seeded with an initial velocity, as well as a communication channel between the particles. Particles then move through the solution space and are evaluated according to some fitness criterion after each time step. Over time, particles are accelerated toward those particles within their communication grouping that have better fitness values. The main advantage of such an approach over other global minimization strategies such as simulated annealing is that the large number of members that make up the particle swarm make the technique impressively resilient to the problem of local minima.

- Stochastic diffusion search

  Stochastic diffusion search (SDS) is an agent based on probabilistic global search and optimization technique best suited to problems where the objective function can be decomposed into multiple independent partial functions. Each agent maintains a hypothesis, which is iteratively tested by evaluating a randomly selected partial objective function parameterized by the agent's current hypothesis. In the standard version of SDS, such partial function evaluations are binary, resulting in each agent becoming active or inactive. Information on hypotheses is diffused across the population via inter agent communication. Unlike the stigmergic communication used in ACO, in SDS agents communicate hypotheses via a one-to-one communication strategy analogous to the tandem running procedure observed in some species of ant. A positive feedback mechanism ensures that, over time, a population of agents stabilize around the global-best solution. SDS is both an efficient and robust search and optimization algorithm, which has been extensively mathematically described.

The use of Swarm Intelligence in Telecommunication Networks has also been researched, in the form of Ant-Based Routing. Basically, this uses a probabilistic routing table rewarding/reinforcing the route successfully traversed by each "ant" (a small control packet) which flood the network. Reinforcement of the route in the forward, reverse direction, and both simultaneously have been researched: backward reinforcement requires a symmetric network and couples the two directions together; forward reinforcement rewards a route before the outcome is known. As the system behaves stochastically and is therefore lacking repeatability, there are large hurdles to commercial deployment. Mobile media and new technologies have the potential to change the threshold for collective action due to swarm intelligence.

Artificial Immune Systems

The field of Artificial Immune Systems (AIS) is concerned with abstracting the structure and function of the immune system to computational systems, and investigating the application of these systems toward solving computational problems from mathematics, engineering, and information technology. AIS is a subfield of computational intelligence, biologically inspired computing, and natural computation, with interests in Machine Learning and belonging to the broader field of artificial intelligence.

Artificial Immune Systems (AIS) are adaptive systems, inspired by theoretical immunology and observed immune functions, principles, and models, which are applied to problem solving.

AIS is distinct from computational immunology and theoretical biology that are concerned with simulating immunology using computational and mathematical models toward better understanding the immune system, although such models initiated the field of AIS and continue to provide a fertile ground for inspiration. Finally, the field of AIS is not concerned with the investigation of the immune system as a substrate computation, such as DNA computing.

The common techniques are inspired by specific immunological theories that explain the function and behavior of the mammalian adaptive immune system.

- Clonal Selection Algorithm: A class of algorithms inspired by the clonal selection theory of acquired immunity that explains how B and T lymphocytes improve their response to antigens over time called affinity maturation. These algorithms focus on the Darwinian attributes of the theory where selection is inspired by the affinity of antigen-antibody interactions, reproduction is inspired by cell division, and variation is inspired by somatic hypermutation. Clonal selection algorithms are most commonly applied to optimization and pattern recognition domains, some of which resemble parallel hill climbing and the genetic algorithm without the recombination operator.

- Negative Selection Algorithm: Inspired by the positive and negative selection processes that occur during the maturation of T cells in the thymus called T cell tolerance. Negative selection refers to the identification and deletion (apoptosis) of self-reacting cells, that is, T cells that may select for and attack self tissues. This class of algorithms are typically used for classification and pattern recognition problem domains where the problem space is modeled in the complement of available knowledge. For example, in the case of an anomaly detection domain the algorithm prepares a set of exemplar pattern detectors trained on normal (nonanomalous) patterns that model and detect unseen or anomalous patterns.

- Immune Network Algorithms: Algorithms inspired by the idiotypic network theory describe the regulation of the immune system by anti-idiotypic antibodies (antibodies that select for other antibodies). This class of algorithms focus on the network graph structures involved where antibodies (or antibody producing cells) represent the nodes and the training algorithm involves growing or pruning edges between the nodes based on affinity (similarity in the problems representation space). Immune network algorithms have been used in clustering, data visualization, control, and optimization domains, and share properties with artificial neural networks.

4.9.2 Bio-inspired Network Management

Following the incredibly high similarity between future networks and biological systems, an increasing number of researchers are trying to draw inspiration from the living world and to apply biological models. Through time biological systems are able to evolve and learn and adapt to changes that environment brings upon living organisms. Researchers try to investigate the various biological processes that exhibit self-governance and combine these different processes into a biological framework that can be applied to various communication systems to realize true autonomic behavior.

The properties of self-organization, evolution, robustness, and resilience are already present in biological systems. This indicates that similar approaches may be taken to manage different complex networks, which allows the expertise from biological systems to be used to define solutions for governing future communication networks.

Future network applications are expected to be autonomous, scalable, adaptive to dynamic network environments, and to be simple to develop and deploy. In order to realize future network applications with such desirable characteristics, various biological systems have already developed the mechanisms necessary to achieve the key requirements of future network applications such as autonomy, scalability, adaptability, and simplicity.

Bio-inspired Framework

The management stage of the communication system will include mechanisms that enable the system to sustain itself. These mechanisms include the ability for the system to balance the internal equilibrium (homeostasis) and rely on the self-organization mechanism for coordination with other elements to further support sustainable management in the face of environmental changes.

Bio-networking architecture is designed based on the three principles described below in order to interact and collectively provide network applications that are autonomous, scalable, adaptive, and simple.

1. Decentralization. Decentralization allows network applications to be scalable and simple by avoiding a single point of performance bottleneck and

failure and by avoiding any central coordination in developing and deploying network applications and services.

2. Autonomy: Bio-network elements monitor their local environments, and based on the monitored environmental conditions, and autonomously interact without any intervention from human users or from other controlling entities.

3. Adaptability: Bio-networks are adaptive to dynamically changing environmental conditions (e.g., user demands, user locations, and resource availability) over the shortterm and longterm.

A bio-inspired architecture for network management is illustrated in Figure 4.16 [BBD+07].

The architecture is composed of various components that are subcategorized under self-organization, self-management, and self-learning.

Network infrastructure requires efficient resource management that must satisfy the business objective of the operator. At the same time, the robustness and scalability issue are a core requirement in network infrastructure. These factors are crucial when supporting changes that require route and resource reconfiguration. This factor may be due to changes in traffic behavior. Based on these characteristics, some researchers provide homeostasis equilibrium model to the self-management of resources, chemotaxis for self-organization of dynamic routes, and evolutionary computation for self-learning to support dynamic traffic changes.

Autonomic network management provides a communication system with the ability to manage and control complexity in order to enhance self-governance. Figure 4.17 [BBD+07] presents an application architecture for self-management. It is based on three-layer hierarchy: Business, Systems, and Device layers. The business level codifies business goals, and translates them into a form that can be used by the system and device layers. As presented in Figure 4.17, homeostasis model is used for resource management, while the self-organization mechanism for decentralized routing, the chemotaxis, reaction diffusion, and hormone signaling model are employed.

Bio-inspired strategies is helpful to approach the goal of autonomic network management and self-management. An example research of bio-inspired approach for autonomic route management can be find in [BBJ+09].

4.10   XML in Network Management

4.10.1   Introduction of XML

The Extensible Markup Language (XML) is a general-purpose specification for creating custom markup languages. It is classified as an extensible language, because it allows the user to define the mark-up elements. XML's purpose is to aid information systems in sharing structured data, especially via the Internet, to encode documents, and to serialize data; in the last context, it compares with text-based serialization languages such as JSON and YAML.

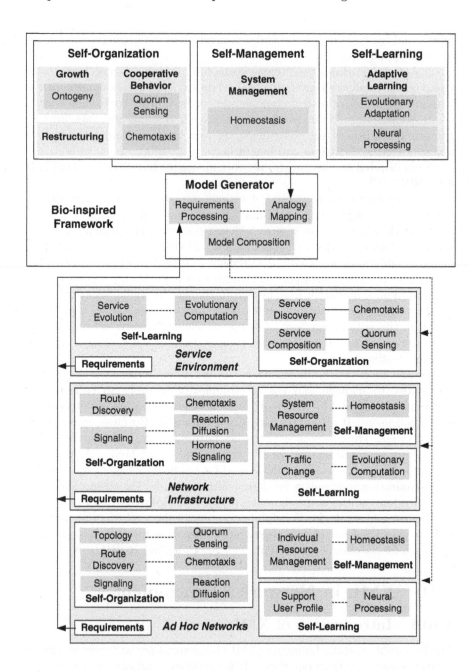

Figure 4.16: Bio-inspired Architecture for Network Management

XML's set of tools help developers in creating web pages, but its usefulness goes well beyond that. XML, in combination with other standards, makes it

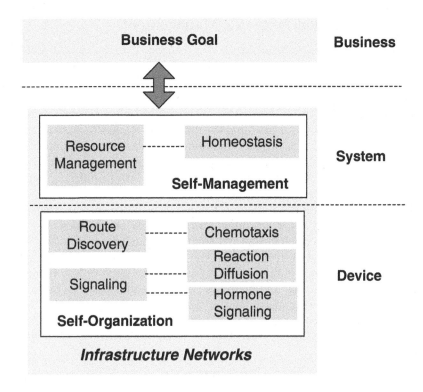

Figure 4.17: Application Architecture of Bio-inspired Self-Management

possible to define the content of a document separately from its formatting, making it easy to reuse that content in other applications or for other presentation environments. Most important, XML provides a basic syntax that can be used to share information between different kinds of computers, different applications, and different organizations without needing to pass through many layers of conversion.

XML began as a simplified subset of the Standard Generalized Markup Language (SGML), meant to be readable by people via semantic constraints; application languages can be implemented in XML. These include XHTML, RSS, MathML, GraphML, Scalable Vector Graphics, MusicXML, and others. Moreover, XML is sometimes used as the specification language for such application languages.

XML supports several standards such as XML Schema, document object model (DOM) application program interface (API), XML path language (XPath), Extensible Stylesheet Language (XSL), XSL transformations (XSLT), etc. Many efforts are in progress to apply XML technologies and their previously existing implementations to a wide range of network management.

4.10.2 XML-Based Network Management

XML-based network management uses XML to encode communications data, providing an excellent mechanism for transmitting the complex data that are used to manage networking gear.

Building an API around an XML-based remote procedure call (RPC) gives a simple, extensible way to exchange this data with a device. Receiving data in XML opens easy options for handling the data using standards-based tools. XML is widely accepted in other problem domains, and free and commercial tools are emerging at an accelerating rate. Using extensible stylesheet language transformations (XSLT) enables you to recast data in differing lights and convert it to and from common formats.

XML-based RPCs provide a standards-based API to devices, allowing external applications to take full advantage of the devices' capabilities. Use of traditional device access mechanisms and communication protocols allows access to a device under any circumstance. Use of existing authentication and authorization mechanisms minimizes the amount of new code in use, reduces the risk of security problems, and increases the trust and comfort of users.

Using XML in network management presents many advantages:

- The XML Schema defines the structure of management information in a flexible manner.

- Widely deployed protocols such as HTTP reliably transfer management data.

- DOM APIs easily access and manipulate management data from applications.

- XPath expressions efficiently address the objects within management data documents.

- XSL processes management data easily and generates HTML documents for a variety of user interface views.

- Web Service Description Language (WSDL) and simple object access protocol (SOAP)define web services for powerful high-level management operations.

Four possible combinations between managers and agents can be considered for XML-based integrated network management (see Figure 4.18).

The example architecture of an XML-based manager, an XML-based agent, and an XML/SNMP gateway for XML-based integrated network management can be designed as Figures 4.19, 4.20, 4.21 [CHJ03].

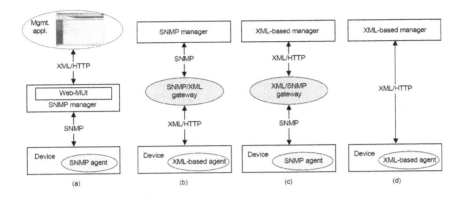

Figure 4.18: XML-based Combinations of Manager and Agent

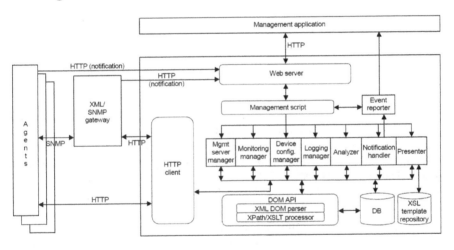

Figure 4.19: Architecture of an XML-Based Manager

4.11 Other Techniques for Network Management

4.11.1 Economic Theory

Network management using economic theory proposes to model network services as an open-market model. The resulting network is self-regulating and self-adjusting, without the presence of any formal network management infrastructure. Network administrators would indirectly control the systems dynamics by inducing incentives and define aggregate economic policies. Such an approach may seem to be very bold, but it draws its theory from the well-

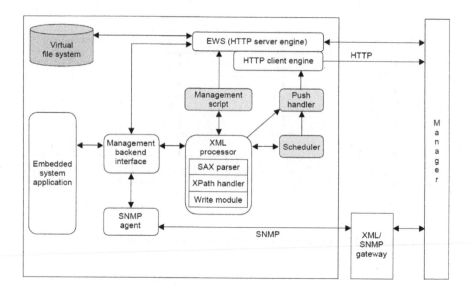

Figure 4.20:   Architecture of an XML-Based Agent

established economic sciences. The premises for applying economic theories are: the existence of open and heterogeneous networks; multivendor orientation; and competitive services. Very little work has been done on this subject, and most of it is focused on using economic theory as an agent coordination model [BKR99] [BMI+00].

However, the application of economic theories to network management is only at an early experimental stage. Many critical issues brought out with these experiments cast doubts on the applicability of economic theory to network management. Using a market model for managing networks is a novel idea. However, some important design issues must be carefully considered.

- Firstly, the driving force for a market model is the authenticity of its currency. The currency in network management denotes the authority to take use of system resources for certain component or certain service. Hence, currency values and their transaction processes used in a market model must be secure. Furthermore, such secure transactions must be performed very efficiently, as it would be a very frequent operation.

- Secondly, the economic policy for the market model must be designed in such a way that it encourages fair competition, and strongly relates resource contention and its associated price.

- Lastly, the market model would be operating on a wide scale, requiring standardization of its elements and operations. Such standardization

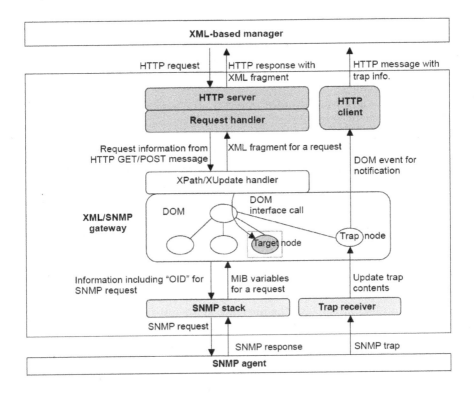

Figure 4.21: Architecture of an XML/SNMP Gateway

may be a very slow process and would require full consensus from all participating vendors.

4.11.2 Finite-State Machines

A finite-state machine (FSM) is an abstract model describing a synchronous sequential machine and its spatial counterpart, the interactive network [Koh78].

In [WS93a] the authors address the problem of fault detection, which is the first step in fault management. Given a system that can be modeled by an FSM, its behavior can be described by a word, w, which is a concatenation events. The fault detection problem can be defined as the difference between normal and faulty words. Faults are modeled as abnormal words that violate the specification. A system under observation is given by an FSM that is a 4-tuple $G = (\Sigma, Q, \delta, \varrho)$, where Σ is the set of all possible events or the alphabet. Q is a finite nonempty set of states. δ is the state transition function: $Q \times \Sigma \to Q$. ϱ is the initial state set of the FSM. $L(G)$ denotes the language generated by FSM $G \in Q$.

The goal of [WS93a] is to build an observer, i.e., a machine A, recognizes that faulty words violating the specification. Since $L(G)$ is the set of correct words generated from G, $\overline{L(G)}$ denotes the set of faulty words. There are several objectives in the choices of fault detection systems. One is to minimize the information fed to the observer to detect a given class of faults. Another is to optimize the structure of the observer. However, it is inevitable that not all faults can be detected if the complexity of the observer is simpler than the original system.

In [BHS93] the authors investigate the problem of fault identification in communication networks. Fault identification is the process of analyzing the data or output from the malfunctioning component in order to propose possible hypotheses for the faults. Therefore, this work assumes that fault localization methods have already isolated a fault in a single process of a communication system. The faulty process is modeled as an FSM, and the possible faults are represented as additions and changes to the arcs in that FSM. The trace history of the faulty process is used in order to propose possible hypotheses about multiple faults. The trace history is assumed to be partially observed (i.e., it may include deletions, additions, and changes of events). Therefore, the problem becomes one of the inferring finite state structures given by the unreliable and partially observed trace history.

[RH95] presents the problem of alarm correlation using probabilistic FSM (PFSM). A fault is modeled as a PFSM so that the possible output sequences correspond to the possible alarm sequences resulting from the fault. The data model includes a noise description, which allows the machine to handle noisy data. Therefore, this approach can be divided into two phases: a learning phase, which acquires the model of each fault from possibly incomplete or incorrect history data, and a correlation phase, which uses those models to interpret the online data and identify faults. Both phases call for heuristic algorithms.

This approach can be viewed as a centralized approach where the network is considered a single entity. Here, the events are collected in a centralized location and then correlated with each other. This approach has difficulty in scaling up to large networks.

4.11.3 Model-Traversing Techniques

Model-traversing techniques use a formal representation of a network with clearly marked relationships among system entities [JP93] [Kat96] [KP97]. By exploring these relationships, starting from the system entity that reported an alarm, the fault identification process is able to determine which alarms are correlated and locate faulty elements of the networks.

Model-traversing techniques reported in the literature use an object-oriented representation of the system [HCF95]. One approach [JP93] exploits the OSI management framework. The approach described in [Kat96] uses guidelines for the definition of managed objects (GDMO) (with nonstandard extensions) to model services and dependencies between services in a network. The proposed refinements of this model include the possibility of testing operational and quality of service states of managed services [KG97].

During the model traversal, managed objects are tested to determine their operational status. The root cause is found when the currently explored malfunctioning object does not depend on any other malfunctioning object. In multilayer models, first a horizontal search is performed in the layer in which the failure has been reported [KP97]. When a failing component is located, a vertical search carries the fault localization process to the next layer down. In the lower layer, the horizontal search is started again. In NetFACT [HCF95], the fault localization process is performed in two phases. Firstly, in the horizontal search, votes are assigned to the potentially faulty elements based on the number of symptoms pertaining to these elements. Secondly, in the phase-tree search, the root cause is determined. This search determines whether the device that receives the most votes in the first step is at fault or whether it fails because one of the components it depends upon is faulty.

Model-traversing techniques are robust against the frequent configuration changes in networks [KP97]. They are particularly attractive when the automatic testing of a managed object may be done as a part of the fault localization process. Model-traversing techniques seem natural when relationships between objects are graph-like and easy to obtain. These models naturally enable the design of distributed fault localization algorithms. Their strengths are high performance, potential parallelism, and robustness against a network's changes. Their disadvantage is a lack of flexibility, especially if fault propagation is complex and not well structured. In particular, they are unable to model situations in which the failure of a device may depend on a logical combination of other device failures [HCF95].

Chapter Review

1. How can policy-based approach be used for network management?

2. Which kinds of network management tasks can be resolved by artificial intelligence approaches?

3. How do graph-theoretic techniques model dynamic topology networks?

4. What are the advantages of probabilistic approaches for network management?

5. What are the advantages and disadvantages of agent techniques for network management?

6. What's the difference between policy-based approach and active network technique for network management?

7. What are the advantages and challenges of bio-inspired network management?

8. How can the web-based approach and XML improve network management?

9. To realize autonomic network management, which technologies could be candidates?

10. Investigate some approaches to resolve an example of emerging network management challenges.

Chapter 5

Management of Emerging Networks and Services

5.1 Next Generation Networking

5.1.1 Introduction of Next Generation Networking

Next Generation Networking (NGN) is a broad term to describe some key architectural evolutions in telecommunication core and access networks that will be deployed over the next 5-10 years. The general idea behind NGN is that one network transports all information and services (voice, data, and all sorts of media such as video) by encapsulating these into packets, like it is on the Internet. NGNs are commonly built around the Internet Protocol, and therefore the term "all-IP" is sometimes used to describe the transformation toward NGN.

It is clear that the NGN is not just another network, which will be deployed and operated next to, and independent from the existing telecommunication network. Instead, over time, parts of the existing network will be replaced by NGN structures, and other parts of the existing network will be integrated into the NGN. For example, the existing 64 kbit/s circuit switched voice network (PSTN) will completely disappear over time. Its function will be absorbed by the universal packet transfer and routing capability of the NGN. This capability will support the routing of packet streams of widely varying capacity, carrying not only voice but also other numerous other forms of information. Actually, the architecture of the NGN supports the view that voice transport will rapidly become one of the smaller applications of the network. On the other hand, access networks based on xDSL technologies will probably not disappear but become an integral part of the NGN, next to other access wireline and wireless access technologies.

ITU-T Rec. Y.2012 describes a number of NGN concepts, and it provides the specification of the NGN components and the functional architecture. Figure 5.1 presents the overview of NGN Architecture [ITU2012].

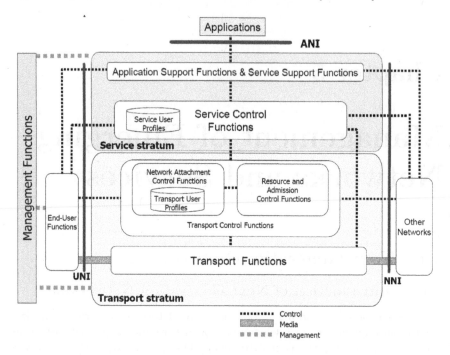

Figure 5.1: NGN Architecture Overview

Some main architectural aspects of the NGN (and its environment) that are shown in this figure are:

- the two separate strata, which are the essential elements of an NGN, i.e., the Service Stratum and the Transport Stratum;

- the Applications, which reside outside the NGN;

- the End User Functions and the Other Networks to which the NGN is connected;

- the Management Functions that manage the NGN.

The NGN Service Stratum provides the functions that control and manage network services to enable the end-users services and applications. End-user services may be implemented by a recursion of multiple service strata within the network. Services may be related to voice, data, or video applications, arranged separately or in some combination in the case of multimedia applications.

The NGN Transport Stratum is concerned with the transfer of information between peer entities. For the purposes of such transfers, dynamic or static associations may be established to control the information transfer between such entities. Associations may be of durations that are extremely short, medium term (minutes), or long term (hours, days, or longer).

The main purpose of the separation of applications, the transport stratum and the service stratum is to allow independent evolution of the technologies used in these strata. For example, in the current view, the IP Multimedia Subsystem (IMS), specified by the 3rd Generation Partnership Project (3GPP), is an important candidate for the service control functions. If ever a newer technology would replace IMS, the separation of the strata should allow this evolution without affecting the transport stratum. Conversely, the same is true for the other strata: one of the technologies for the transport stratum is T-MPLS over SDH. Again, this technology may be replaced by others without affecting the service control layer or the applications supported by the network.

IP Reference Model for NGN

The IP reference model and architecture will support the implementation of NGN services, and verify Net Centricity across an all IP network. Existing circuit-based non-IP networks, connected to all IP networks, for an interim period will need to identify transformation components to interoperate with the architecture of the IP Reference Model. The NGN services planes will reside within the IP reference model Applications Layer and provide low-level interfaces to pass information to and from the communications layers.

The Figure 5.2 depicts the NGN service delivery plane, network application layer, and session and stream control layer. It is an architecture model where the software to deliver services ar depicted as a service framework.

The Network Application Layer and the Services Deliver Plane provide the capabilities to develop the services required by NGN. The User and Services Plane provide NGN services over the Platform OS and Network Infrastructure communications layer.

It is also a method for the Services Framework to affect, through network configuration parameters, the behavior of the communications layers, which is the IP Reference Model Effect from NGN. This alters the way IP networks today are managed, constructed, and secured.

Thus, there are two effects to the IP Reference Model from NGN.

- The first is that the Applications Layer will become a Services Framework supporting many sublayers depending on the services required, and help address the Technology Challenges discussed.

- The second is that the network communications layers will use the Services Framework operational parameters to influence the end-to-end communications connectivity.

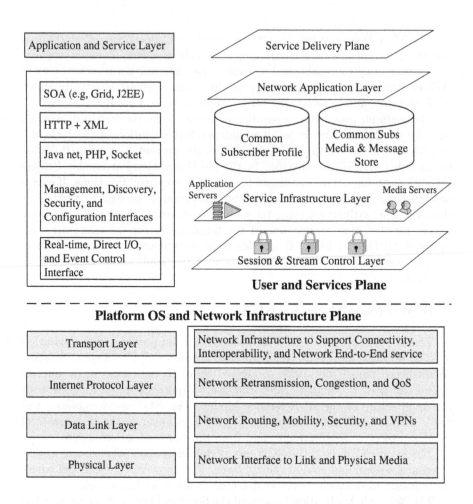

Figure 5.2: NGN Components and Internet Protocol Reference Model Layers

This effect will actually permit a stateless model for the communications layer and support specific services models across an end-to-end IP network. It will inject and receive network knowledge to and from the Transport, IP, Data Link, and Physical IP Reference Model Layers.

The IP Multimedia Subsystem (IMS) is a standardized NGN architecture for an Internet media-services capability defined by the European Telecommunications Standards Institute (ETSI) and the 3rd Generation Partnership Project (3GPP).

The realization of the NGN will require protocol extensions within each layer of the Internet Protocol reference model: e.g., IPv6, SCTP, IPsec, SIP, RTP, and RTSP. These extensions need to be included in every relevant NGN solution offering. The implementation result and infrastructure to support this NGN Effect will vary depending on node type (e.g., Server, Client, Router, Handheld, or Sensor).

Figure 5.3 presents an example of NGN Realization.

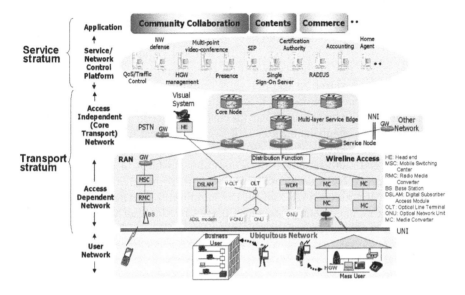

Figure 5.3:   Example of NGN Realization

5.1.2   Management of Next Generation Networking

Management Challenges for NGN

The deregulation, competition, and rapid technology development for NGN generates significant challenges in terms of operation, administration and maintenance of networks and services [LS05].

- Dynamic Topology

  In NGN, it is reasonable to expect that devices, especially high-end routers and switches, will become increasingly programmable, and that it will become possible to execute more control software directly on the devices. As a result, network topology of common networks of NGN can change occasionally. In addition, the collaboration between disparate network domains or between different service providers will increase to a great extent. Dynamic configuration and topology of NGN will challenge the traditional configuration management approaches, which are often inefficient and involve too many human efforts. In NGN, a quick-response and network-wide configuration capability is required to manage the changing network topology that may be composed of thousands of distributed nodes.

- Heterogeneity

  The NGN will not only contain the legacy components from traditional PSTN, but also some "brand new" components from the development of up-to-date technologies, e.g., Multi-protocol Label Switching (MPLS). Meanwhile, the flexibility based on trust negotiation among disparate domains is required in the pervasive computing environments of NGN. As the Internet has proven, it is impractical for a single service provider to roll out all the services that its customers need. The interoperability among heterogeneous entities will become critical important for NGN. For these reasons, different vendors' platforms/technologies have to be "converged" and managed on a common platform in order to support and improve NGN services. Together with some emerging approaches, both CMIP and SNMP can be the candidates for the next-generation network management protocols. Limited by the multivendor capability and other weaknesses of current approaches, how to deal with heterogeneous resources in a cost-effective manner thus becomes the big challenge for NGN.

- Multiple Services

  The NGN is packet-based and responsible for carrying multiple services over the single IP-based transport network, ranging from traditional telephony voice to data, video and multimedia applications. Apart from the best-effort approach in the current Internet, the NGN is optimized for differentiated services where QoS and reliability of services will be engineered and guaranteed. Accordingly, the traffic management capability for differentiated NGN services and traffic has to be provided so as to monitor and control any concerned service. In the traditional TMN framework, traffic management has not been addressed clearly since all network connections are at fixed rate. In NGN, the fine-grained controlling and monitoring of traffic pattern will become an important consideration for NGN service providers and network operators.

- Standardization

  For any service provider or network operator in NGN, the biggest motivation for adopting new operations support system (OSS) is to maximize Return On Investment (ROI). Besides taking advantage of new technologies coping with issues such as multiple services, other industry trends have to be considered, such as the trend toward commercial off-the-shelf (COTS) components and systems promising seamless integration (plug-and-play). Most important of all, the fundamental management architecture for NGN shall be considered. In the TMN architecture of the ITU-T, no further decomposition of the proposed layers into specific functions is proposed. A universally agreed set of management requirements for NGN are lacked. In order to support the core functions of NGN, the management framework, architecture, information model and management protocols have to be standardized and agreed among a number of NGN participants. Although ITU-T's NGN Management Focus Group is emerging for necessary management standards, standardization in the area of network management for NGN is still fragmented at many different standards bodies.

Management Architecture for NGN

NGN Management functionality is based on the ITU-T Rec. M.3050 series, Enhanced Telecom Operations Map (eTOM). Originated by the TMForum, the eTOM identifies the processes required to run a telecoms network. In successive analytical steps, the identified processes are split up into subprocesses, until a level of detail is obtained that is sufficient to enable a meaningful exchange of views between management product vendors and network operators. The outcome of the second step is reflected in an set of so called Level 1 processes. Figure 5.4 shows these processes in the context of the overall map.

The process taxonomy as developed in the eTOM does not necessarily lead to a meaningful physical management systems architecture. Some of the processes may be implemented by human resources and other may be spread out of a number of systems. Yet, the identification of relatively self-contained processes helps in identifying functional interfaces. This is an important aspect as it allows the NGN management functionality to be specified and implemented based on the principles of a Service-Oriented Architecture (SOA).

One of the architectural principles behind the management architecture for Next Generation Networks is that of being a Service-Oriented Architecture.

- Service-Oriented Architecture for Management of NGN

  A Service-Oriented Architecture (SOA) is a software architecture of services, policies, practices, and frameworks in which components can be reused and repurposed rapidly in order to achieve shared and new functionality. This enables rapid and economical implementation in response to new requirements thus ensuring that services respond to perceived user needs.

  SOA uses the object-oriented principle of encapsulation in which entities are accessible only through interfaces and where those entities are connected by well-defined interface agreements or contracts.

  Major goals of an SOA in comparison with other architectures used in the past are to enable:

  - rapid adaptation to changing business needs;

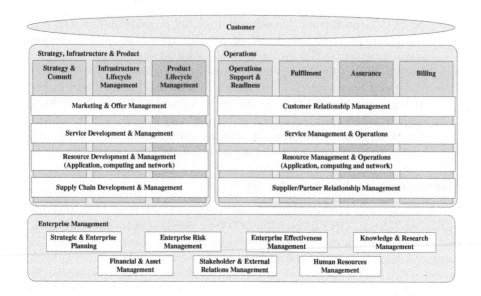

Figure 5.4:   eTOM Business Process Framework

— cost reduction in the integration of new services, as well as in the maintenance of existing services.

SOA supports the generation of open and agile business solutions that can be rapidly extended or changed on demand. This will enable NGN Management to support the rapid creation of new NGN services and changes in NGN technology.

The main features of SOA are:

— it is loosely coupled, location independent, and supports reusable services;

— any given service may assume a consumer and provider role with respect to another service, depending on the situation;

— the "find-bind-execute" paradigm for the communication between services;

— published contract-based, platform, and technology-neutral service interfaces. This means that the interface of a service is independent of its implementation;

– encapsulating the life cycle of the entities involved in a business transaction; and exposing a coarser granularity of interfaces than an Object-Oriented Architecture.

Web services are promising in supporting a wide range of distributed services and transparently hiding technical implementation details, and thereby reduce integration costs.

Web services are composed of several building blocks built on top of XML and SOAP, the latter of which is a stateless message exchange mechanism. Because SOAP messages can work with standard Web protocols, such as XML, HTTP, and TCP/IP, they can function well across heterogeneous network components. The ubiquitous availability of HTTP and the simplicity of XML-based SOAP make web services ideal for system interconnections. More details about web-based network management and XML-based network management can be found in Section 4.5 and Section 4.10.

- Standard Management for NGN

  Management functionality may also be defined in terms of the descriptions of what are termed function sets in support of Element, Network, Service, and Business Management defined in ITU-T Rec. M.3400. These function set descriptions are categorized according to fault, configuration, accounting, performance, and security (FCAPS) management and populated with functional descriptions that mirror the traditional division of telecom management functionality into Operations, Administration, Maintenance, and Provisioning (OAM&P).

  The relation between the traditional FCAPS management functions and the eTOM processes is described in Supplement 3 to ITU-T Rec. 3050.

  Management of a NGN environment is an information processing application. To effectively manage complex networks and support network operator/service provider business processes, it is necessary for management applications to interact by directing/performing actions and exchanging information implemented in multiple consumer and provider entities. Thus telecommunication management is a distributed application. In order to promote interoperability, NGN management is based on standardized, open management paradigms that support the standardized modeling of the information to be communicated across interfaces between consumer entities and provider entities. Management standardization activities generally do not develop a specific management paradigm but build upon industry-recognized solutions, focusing primarily on object-oriented and service-oriented techniques. Specific management paradigms and information architectural principles may be applied in management standards when judged to be adequate.

In general, the management requirements for the solution need to be stated in business terms and to be decomposed to identify application interactions in business terms sufficient to drive the definition of information models and interface operations; the information model details must be traceable to requirement details; information must be specified independent of implementation technology, i.e., in a protocol-neutral form using industrial strength methods and tools, and also specified in a protocol-specific form traceable to the protocol-neutral form; and communicated using standardized transport mechanisms.

5.2 Wireless Networks

5.2.1 Advances in Wireless Networks

Wireless network refers to any type of computer network that is wireless and is commonly associated with a telecommunications network whose interconnections between nodes is implemented without the use of wires. Wireless telecommunications networks are generally implemented with some type of remote information transmission system that uses electromagnetic waves, such as radio waves, for the carrier and this implementation usually takes place at the physical level or "layer" of the network.

With the evolution of technologies and application in recently years, wireless networks evolved to different type of networks. The marriage of the Internet and the wireless technologies yields the IP-based wireless networking.

Figure 5.5 presents a classification wireless networks.

The varying wireless network technologies are depicted in Figure 5.6 together with the corresponding IEEE standards.

Figure 5.7 [KF07] illustrates the current mosaic of wireless communication networks from the service coverage (range) standpoint. Two main network components are clearly distinguished, namely, wide area networks on one hand, and short-range networks on the other hand. Curiously, range-wise, the development of wireless communication networks follows an ordered evolution from large to small networks, starting with very large distribution networks of up to hundred of kilometers wide down to submeter short-range networks. Several reasons can be attributed to the development of increasingly smaller wireless networks, including the pressure to move toward unused (and typically higher) frequency bands of the spectrum and the need to support higher data throughputs. In general these two component networks were developed independently of each other but aiming, by design, to coexist.

Short-range wireless communications involve a very diverse array of air interface technologies, network architectures, and standards. The most well known short-range wireless network technologies include wireless local area networks (WLAN), wireless personal area networks (WPAN), wireless body area networks (WBAN), wireless sensor networks (WSN) car-to-car

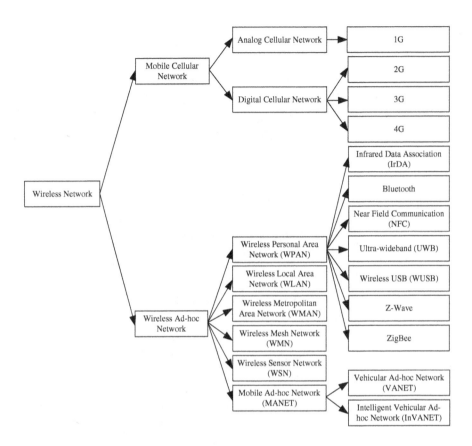

Figure 5.5: Classification of Wireless Networks

communications (C2C), Radio Frequency Identification (RFID), and Near Field Communications (NFC). An overview of short-range communications from the standpoint of the typical range is presented in Figure 5.8 [Dav08]. As compared to long-range communications, short-range links require significantly lower energy per bits in order to establish a reliable link. They can achieve thus data throughputs of several orders of magnitude higher than the typical values for cellular networks. Usually short-range networks exploit distributed architectures, using unlicensed spectrum.

Short-range communications encompass a large variety of different wireless systems with a great diversity of requirements. Figure 5.9 [Dav08] illustrates the wide scope of short-range communications by classifying it according to the most common air interface technologies and network architectures, as well as supported mobility and data rates.

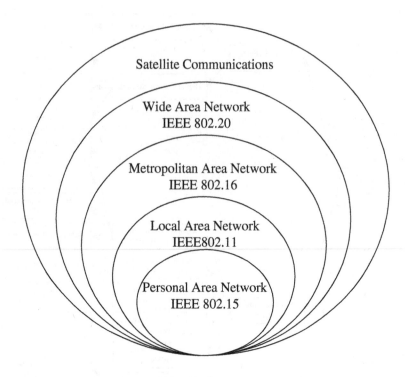

Figure 5.6: Wireless Networks Technologies Based on IEEE Standards

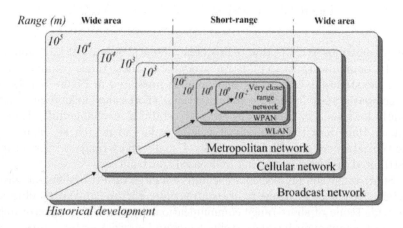

Figure 5.7: Future Wireless Communications: An All – Encompassing Network of Networks

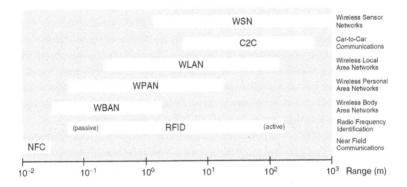

Figure 5.8:   A Classification of Short-Range Communications According to the Typical Supported Range

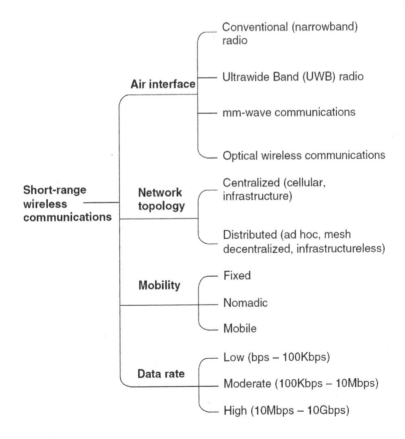

Figure 5.9:   A General Classification of Short-range Communications

5.2.2   Mobile Cellular Networks

Mobile cellular networks are to transfer information over a distance without the use of electrical conductors or "wires." The distances involved may be short (a few meters as in television remote control) or very long (thousands or even millions of kilometers for radio communications). When the context is clear the term is often simply shortened to "wireless." Wireless communication is generally considered to be a branch of telecommunications.

It encompasses various types of fixed, mobile, and portable two way radios, cellular telephones, personal digital assistants (PDAs), and wireless networking. Other examples of wireless technology include GPS units, garage door openers, and garage doors, wireless computer mice, keyboards and headsets, satellite television, and cordless telephones.

A mobile phone (also known as a wireless phone, cell phone, or cellular telephone) is a short-range, electronic device used for mobile voice or data communication over a network of specialized base stations known as cell sites. In addition to the standard voice function of a mobile phone, telephone, current mobile phones may support many additional services, and accessories, such as SMS for text messaging, e-mail, packet switching for access to the Internet, gaming, Bluetooth, infrared, camera with video recorder, and MMS for sending and receiving photos and video. Most current mobile phones connect to a cellular network of base stations (cell sites), which is in turn interconnected to the public switched telephone network (PSTN) (the exception is satellite phones).

Analog Cellular Networks

The first generation of cellular networks (1G systems) was designed in the late 1960s. These are the analog cell phone standards that were introduced in the 1980s and continued until being replaced by 2G digital cell phones. The 1G systems were composed of mainly analog signals for carrying voice and music. There were one directional broadcast systems such as television broadcast, AM/FM radio, and similar communications. 1G can be thought of as descendants of MTS/IMTS since they were of also analog systems.

Digital Cellular Networks

The second-generation wireless telephone networks (2G) use digital systems for communication. Compared to analog systems, digital systems have a number of advantages:

- Digitized traffic can easily be encrypted in order to provide privacy and security. Encrypted signals cannot be intercepted and overheard by unauthorized parties (at least not without very powerful equipment). Powerful encryption is not possible in analog systems, which most of the time transmit data without any protection. Thus, both conversations and network signaling can be easily intercepted. In fact, this has been a significant problem in 1G systems since in many cases eavesdroppers picked up user's identification numbers and used them illegally to make calls.

- Analog data representation made 1G systems susceptible to interference, leading to a highly variable quality of voice calls. In digital systems, it is possible to apply error detection and error correction techniques to the voice bitstream. These techniques make the transmitted signal more robust, since the receiver can detect and correct bit errors. Thus, these techniques lead to clear signals with little or no corruption, which of course translates into better call qualities. Furthermore, digital data can be compressed, which increases the efficiency of spectrum use.

- In analog systems, each RF carrier is dedicated to a single user, regardless of whether the user is active (speaking) or not (idle within the call). In digital systems, each RF carrier is shared by more than one user, either by using different time slots or different codes per user. Slots or codes are assigned to users only when they have traffic (either voice or data) to send.

1. **2G**

Second-generation 2G cellular telecom networks were commercially launched on the GSM standard in Finland by Radiolinja (now part of Elisa Oyj) in 1991. Three primary benefits of 2G networks over their predecessors were that phone conversations were digitally encrypted, 2G systems were significantly more efficient on the spectrum allowing for far greater mobile phone penetration levels; and 2G introduced data services for mobile, starting with SMS text messages.

After 2G was launched, the previous mobile telephone systems were retrospectively dubbed 1G. While radio signals on 1G networks are analog, and on 2G networks are digital, both systems use digital signaling to connect the radio towers (which listen to the handsets) to the rest of the telephone system.

2G technologies can be divided into TDMA-based and CDMA-based standards depending on the type of multiplexing used. The main 2G standards are:

- GSM (Global System for Mobile communications: TDMA-based), originally from Europe but used in almost all countries on all six inhabited continents (Time Division Multiple Access). Today accounts for over 80% of all subscribers around the world.

- IS-95 aka cdmaOne, (CDMA-based, commonly referred as simply CDMA in the US), used in the Americas and parts of Asia. Today accounts for about 17% of all subscribers globally. Over a dozen

CDMA operators have migrated to GSM including operators in Mexico, India, Australia, and South Korea.

- PDC (TDMA-based), used exclusively in Japan.

- iDEN (TDMA-based), proprietary network used by Nextel in the United States and Telus Mobility in Canada.

- IS-136 aka D-AMPS (TDMA-based, commonly referred as simply TDMA in the US) was once prevalent in the Americas but most have migrated to GSM.

2G services are frequently referred to as Personal Communications Service, or PCS, in the United States.

2. Evolution from 2G to 3G

2G networks were built mainly for voice data and slow transmission. Because of rapid changes in user expectation, they do not meet today's wireless needs. Evolution from 2G to 3G can be subdivided into following phases:

- From 2G to 2.5G (GPRS)

  The first major step in the evolution to 3G occurred with the introduction of General Packet Radio Service (GPRS). So the cellular services combined with GPRS became 2.5G.

  GPRS could provide data rates from 56 kbit/s up to 114 kbit/s. It can be used for services such as Wireless Application Protocol (WAP) access, Short Message Service (SMS), Multimedia Messaging Service (MMS), and for Internet communication services such as e-mail and World Wide Web access. GPRS data transfer is typically charged per megabyte of traffic transferred, while data communication via traditional circuit switching is billed per minute of connection time, independent of whether the user actually is utilizing the capacity or is in an idle state.

  GPRS is a best-effort packet switched service, as opposed to circuit switching, where a certain Quality of Service (QoS) is guaranteed during the connection for nonmobile users. It provides moderate speed data transfer, by using unused Time division multiple access (TDMA) channels. Originally, there was some thought to extend GPRS to cover other standards, but instead those networks are being converted to use the GSM standard, so that GSM is the only kind of network where GPRS is in use. GPRS is integrated into GSM Release 97 and newer releases. It was originally standardized by European Telecommunications Standards Institute (ETSI), but now by the 3rd Generation Partnership Project (3GPP).

  2.5G is a stepping stone between 2G and 3G cellular wireless technologies. It does not necessarily provide faster services because bundling of timeslots is used for circuit switched data services (HSCSD) as well. While the terms "2G" and "3G" are officially defined, "2.5G" is not. It was invented for marketing purposes only.

- From 2.5G to 2.75G

  GPRS networks evolved to EDGE networks with the introduction of 8PSK encoding. Enhanced Data rates for GSM Evolution (EDGE), Enhanced GPRS (EGPRS), or IMT Single Carrier (IMT-SC) is a backward-compatible digital mobile phone technology that allows improved data transmission rates, as an extension on top of standard GSM. EDGE can be considered a 3G radio technology and is part of ITU's 3G definition but is most frequently referred to as 2.75G. EDGE was deployed on GSM networks beginning in 2003, initially by Cingular (now AT&T) in the United States.

  EDGE is standardized by 3GPP as part of the GSM family, and it is an upgrade that provides a potential three fold increase in capacity of GSM/GPRS networks. The specification achieves higher data-rates by switching to more sophisticated methods of coding (8PSK), within existing GSM timeslots.

  EDGE can be used for any packet switched application, such as an Internet, video, and other multimedia.

- From 2.75G to 3G

  From EDGE networks the introduction of UMTS networks and technology is called pure 3G.

3. **3G**

3G is the third generation of telecommunication standards and technology for mobile networking, superseding 2.5G. It is based on the International Telecommunication Union (ITU) family of standards under the IMT-2000. Defined by ITU, 3G systems must provide: (1) Backward compatibility with 2G systems; (2) Multimedia support; (3) Improved system capacity compared to 2G and 2.5G cellular systems; and (4) High-speed packet data services ranging from 144 kbps in wide-area mobile environments to 2 Mbps in fixed or in-building environments.

3G networks enable network operators to offer users a wider range of more advanced services while achieving greater network capacity through improved spectral efficiency. Services include wide-area wireless voice telephony, video calls, and broadband wireless data, all in a mobile environment. Additional features also include HSPA data transmission capabilities able to deliver speeds up to 14.4 Mbit/s on the downlink and 5.8 Mbit/s on the uplink.

Unlike IEEE 802.11 networks, which are commonly called Wi-Fi or WLAN networks, 3G networks are wide-area cellular telephone networks that evolved to incorporate high-speed Internet access and video tele-phony. IEEE 802.11 networks are short range, high-bandwidth networks primarily developed for data.

The technical complexities of a 3G phone or handset depends on its need to roam onto legacy 2G networks. In the first country, Japan, there was no need to include roaming capabilities to older networks such as GSM, so 3G phones were small and lightweight. In most other countries, the manufacturers and network operators wanted multi mode 3G phones, which would operate on 3G and 2G networks (e.g., W-CDMA and GSM), which added to the complexity, size, weight, and cost of the handset. As a result, early European W-CDMA phones were significantly larger and heavier than comparable Japanese W-CDMA phones.

The general trend to smaller and smaller phones seems to have paused, perhaps even turned, with the capability of large-screen phones to pro-vide more video, gaming, and Internet use on the 3G networks, and further fueled by the appeal of the Apple iPhone.

The International Telecommunication Union (ITU) defined the demands for 3G mobile networks with the IMT-2000 standard. An organiza-tion called 3rd Generation Partnership Project (3GPP) has continued that work by defining a mobile system that fulfills the IMT-2000 stan-dard. This system is called Universal Mobile Telecommunications Sys-tem (UMTS). Currently, the most common form of UMTS uses W-CDMA as the underlying air interface.

Unlike GSM, UMTS is based on layered services. At the top is the services layer, which provides fast deployment of services and central-ized location. In the middle is the control layer, which helps upgrading procedures and allows the capacity of the network to be dynamically allocated. At the bottom is the connectivity layer where any trans-mission technology can be used and the voice traffic will transfer over ATM/AAL2 or IP/RTP.

4. **4G**

4G (also known as Beyond 3G), an abbreviation for fourth generation, is a term used to describe the next complete evolution in wireless com-munications. A 4G system will be able to provide a comprehensive IP solution where voice, data, and streamed multimedia can be given to users on an "Anytime, Anywhere" basis, and at higher data rates than previous generations.

As 2G was a total replacement of the first-generation networks and handsets; and 3G was a total replacement of 2G networks and hand-sets; so too the 4G cannot be an incremental evolution of current 3G technologies, but rather the total replacement of the current 3G net-works and handsets. The international telecommunications regulatory and standardization bodies are working for commercial deployment of 4G networks roughly in the 2012 – 2015 time scale. At that point, it is predicted that even with current evolutions of third-generation 3G networks, these will tend to be congested.

There is no formal definition for what 4G is; however, there are certain objectives that are projected for 4G. These objectives include that 4G will be a fully IP-based integrated system. 4G will be capable of providing between 100 Mbit/s and 1 Gbit/s speeds both indoors and outdoors, with premium quality and high security.

Many companies have taken self-serving definitions and distortions about 4G to suggest they have 4G already in existence today, such as several early trials and launches of WiMAX. Other companies have made prototype systems calling those 4G. While it is possible that some currently demonstrated technologies may become part of 4G, until the 4G standard or standards have been defined, it is impossible for any company currently to provide with any certainty wireless solutions that could be called 4G cellular networks that would conform to the eventual international standards for 4G. These confusing statements around "existing" 4G have served to confuse investors and analysts about the wireless industry.

4G is being developed to accommodate the quality of service (QoS) and rate requirements set by forthcoming applications like wireless broadband access, Multimedia Messaging Service (MMS), video chat, mobile TV, HDTV content, Digital Video Broadcasting (DVB), minimal service like voice and data, and other streaming services for "anytime-anywhere."

A vision for 4G wireless communications in terms of mobility support and data transmission rate is shown in Figure 5.10 [CG06].

5.2.3 Management of Mobile Cellular Networks

A mobile cellular network is a special kind of wireless system whose features are the following:

- Frequency reuse: The whole coverage area is divided into several smaller areas, called cells, in such a way that some transmission frequencies are used across a set of cells, and reused for another set of cells with little potential for interference.

- Mobility/Roaming: Subscribers are able to move freely around their home network and from this to another one. This feature requires that the network tracks the location of each subscriber in an accurate way, in order to deliver calls and messages properly.

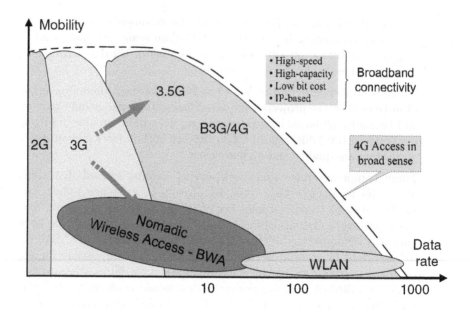

Figure 5.10:   A Vision for 4G Wireless Communications in Terms of Mobility Support and Data Transmission Rate

- Handoff/Handover: The subscriber transitions from one radio channel to another as he/she moves from one cell to another while engaged in a conversation.

Compared to wired networks, wireless cellular networks have some limitations:

- Open Wireless Access Medium: Since the communication is on the wireless channel, there is no physical barrier that can separate an attacker from the network.

- Limited Bandwidth: Although wireless bandwidth is increasing continuously, because of channel contention everyone has to share the medium.

- System Complexity: Wireless systems are more complex due to the need to support mobility and making use of the channel effectively. By adding more complexity to systems, potentially new security vulnerabilities can be introduced.

- Limited Power: Wireless Systems consume a lot of power and therefore have a limited time battery life.

- Limited Processing Power: The processors installed on the wireless devices are increasing in power, but still they are not powerful enough to carry out intensive processing. Relatively Unreliable Network Connection: The wireless medium is an unreliable medium with a high rate of errors compared to a wired network.

These limitations bring more challenges in management of mobile cellular networks:

- Mobility Management:

  Mobility Management is one of the major functions of a GSM or a UMTS network that allows mobile phones to work. The aim of mobility management is to track where the subscribers are, so that calls, SMS, and other mobile phone services can be delivered to them.

  A cellular network is a radio network of individual cells, known as base stations. Each base station covers a small geographical area which is part of a uniquely identified location area. By integrating the coverage of each of these base stations, a cellular network provides a radio coverage over a very much wider area. A group of base stations is called a location area, or a routing area.

  The location update procedure allows a mobile device to inform the cellular network whenever it moves from one location area to the next. Mobiles are responsible for detecting location area codes. When a mobile finds that the location area code is different from its last update, it performs another update by sending to the network, a location update request, together with its previous location, and its Temporary Mobile Subscriber Identity (TMSI).

  There are several reasons why a mobile may provide updated location information to the network. Whenever a mobile is switched on or off, the network may require it to perform an IMSI attach or IMSI detach location update procedure. Also, each mobile is required to regularly report its location at a set time interval using a periodic location update procedure. Whenever a mobile moves from one location area to the next while not on a call, a random location update is required. This is also required of a stationary mobile that reselects coverage from a cell in a different location area, because of signal fade. Thus a subscriber has reliable access to the network and may be reached with a call, while enjoying the freedom of mobility within the whole coverage area.

  When a subscriber is paged in an attempt to deliver a call or SMS and the subscriber does not reply to that page then the subscriber is marked as absent in both the Mobile-services Switching Center (MSC)/Visitor Location Register (VLR) and the Home Location Register (HLR). Mobile not reachable flag (MNRF) is set. The next time the mobile performs a location update the HLR is updated and the mobile not reachable flag is cleared.

The "Temporary Mobile Subscriber Identity" (TMSI) is the identity that is most commonly sent between the mobile and the network. TMSI is randomly assigned by the VLR to every mobile in the area, the moment it is switched on. The number is local to a location area, and so it has to be updated, each time the mobile moves to a new geographical area. A key use of the TMSI is in paging a mobile. "Paging" is the one-to-one communication between the mobile and the base station. The most important use of broadcast information is to set up channels for "paging." Every cellular system has a broadcast mechanism to distribute such information to a plurality of mobiles.

Roaming is one of the fundamental mobility management procedures of all cellular networks. Roaming is defined as the ability for a cellular customer to automatically make and receive voice calls, send and receive data, or access other services, including home data services, when traveling outside the geographical coverage area of the home network, by means of using a visited network. This can be done by using a communication terminal or else just by using the subscriber identity in the visited network. Roaming is technically supported by mobility management, authentication, authorization, and billing procedures.

As the main task of mobility management, location management involves two basic operations:

- Location Updating: Informing the network of a devices location.
- Paging: Polling a group of cells to determine the precise location of a device.

There are three metrics involved with location management: location update cost, paging cost, and paging delay.

The frequency of updates and paging messages relates closely to user movement and call arrival rates, along with network characteristics such as cell size. As mobile devices move between cells in a network they must register their new location to allow the correct forwarding of data. Continual location updates can be a very expensive operation, particularly for users with comparatively low call arrival rates. This update overhead not only puts load on the core (wired) network but also reduces available bandwidth in the mobile spectrum. Importantly, unnecessary location updating incurs heavy costs in power consumption for the mobile device.

An additional area of consideration in location management is the mobility model used to estimate user movement in the network, aiding in the optimization of location updating and paging schemes.

Location update involves reverse control channels while paging involves forward control channels. The total location management cost is the sum of the location update cost and the paging cost. There is a trade-off between the location update cost and the paging cost. If a mobile station updates its location more frequently (incurring higher location update cost), the network knows the location of the mobile station better. Then the paging cost will be lower when an incoming call arrives for the mobile station. Therefore, both location update and paging costs cannot be minimized at the same time. However, the total cost can be minimized or one cost can be minimized by putting a bound on the other cost.

- Resource Management:

Resource management generally aims to guarantee the requested QoS, improve system capacity, and enlarge the coverage of base stations, while keeping cost as low as possible. Resource management includes the congestion control, power and rate allocation, cell planning, and service pricing. Furthermore, the cell planning considers the bandwidth allocation, base station planning, pilot power control, and cell sectorization [ZWX+04].

   - Congestion control

   Congestion control concerns controlling traffic entry into a telecommunications network, so as to avoid congestive collapse by attempting to avoid oversubscription of any of the processing or link capabilities of the intermediate nodes and networks and taking resource reducing steps, such as reducing the rate of sending packets. It should not be confused with flow control, which prevents the sender from overwhelming the receiver.

   Generally speaking, when congestion occurs, three main mechanisms can be applied to relieve the congestion:

   1) to drop some ongoing calls. Then the question is which set of calls to be dropped? The system may choose to drop call(s) in outage condition, or to drop each existing call with prespecified probability, or to drop call(s) that make the largest contribution to alleviate the congestion. The last scheme offers the best performance with the highest complexity;

   2) to decrease transmission rate. This could be done by either proportionally reducing the transmission rate of each user or decreasing the rates to the same maximal fair SIR level;

   3) to reduce the number of simultaneous transmissions. The transmission probability can be dynamically adjusted according to the occupancy information. To eliminate the randomness and to obtain full control over the simultaneous transmissions, cells may sequentially schedule active transmitting periods for each data service.

   - Power and rate allocation.

   Provided with the system parameters such as cell occupancy and channel conditions, it is important to optimize the system performance by efficient allocation of transmission rate and power. A variety of criteria might be optimized, such as to minimize the emitted power, which reduces the battery consumption and causes less interferences; or to maximize transmission rate, which indicates the maximized system throughput and resource utilization.

For the downlink, the objective is to maximize the total number of frames transmitted during a control interval under the SIR (signal-to-interference ratio) constraint for each service type and the average transmission power limit among all base stations. For the uplink, the optimization can be formulated to maximize the total normalized transmission rate subject to the constraint on transmission power and maximum allowed rate. The optimal solution yielded not only higher throughput but also significant power savings. However, this can potentially lead to starvation for users with poor channel condition, since users with better channel condition will always transmit first.

– Cell planning.

In order to reduce the system cost and maximize the utilization of scarce resources, efficient cell planning is of vital importance for service providers. Generally speaking, cell planning consists of bandwidth allocations, base station planning, pilot power control, and cell sectorization.

1) Bandwidth allocations in uplinks and downlinks: To cope with the traffic asymmetry, unbalanced bandwidth allocation is preferred and great system performance gain may be achieved by the proper assignment of bandwidth between two links according to the traffic requirements.

2) Base station planning: To select the sites for base stations by taking into account the system cost, transmission quality, service coverage, and etc. The optimal location of new base stations to minimize a linear combination of installation cost and total transmitted power should take into account traffic distribution, SIR requirements, power allocation constraints, and power control mechanisms. More complex models also exist with considerations on stochastic behavior of the system and/or soft handoff.

3) Pilot power control: Efficient pilot power control needs to balance the cell load and cell coverage area among neighboring cells, whose objectives are to reduce the variation of interference, stabilize the network operation, and improve cell capacity and communication quality especially under nonuniform traffic loading among the cells.

4) Cell sectorization: Adaptive cell sectoring could be deployed at the base station to greatly improve the performance in such a system. It is expected that dynamic cell sectoring could be designed to incorporate the stochastic nature of the system, such as user mo-

bility, channel fading, power control and sectorization imperfection into consideration, to better adapt to the real system.

– Service pricing.

Higher utilization of traffic channels necessitates an efficient reuse. The basic constraint on reuse of radio channels is the interference from the environment or from other mobile devices. Interfering mobile devices may reside in the same cell or may reside in the adjacent cells. Thus, a cell can use only a subset of channels such that interference to a channel by another channel is within the acceptable limit. An efficient Channel Allocation (CA) scheme can reduce these interferences, thereby increasing the channel utilization.

A Dynamic Differentiated Pricing Strategy (DDPS) could be used effectively to regulate demand, whilst offering a sensible means for the joint optimization of network utilization and service provider's revenue. Differentiated service will help a service provider serve its customers economically. In dynamic pricing, tariff varies according to the system's resource utilization. A high tariff rate is fixed during the peak demand period and low tariff rate during the off-peak demand period. This would help the service provider to reduce the idle network capacity during off-peak demand period. As a result, it encourages a more efficient use of available network capacity and improving both the provided QoS and network provider's revenue.

- Security Management:

There are several security issues that have to be taken into consideration when deploying a cellular infrastructure.

– Authentication: Cellular networks have a large number of subscribers, and each has to be authenticated to ensure the right people are using the network. Since the purpose of 3G is to enable people to communicate from anywhere in the world, the issue of cross region and cross provider authentication becomes an issue.

– Integrity: With services such as SMS, chat, and file transfer it is important that the data arrives without any modifications.

– Confidentiality: With the increased use of cellular phones in sensitive communication, there is a need for a secure channel in order to transmit information.

– Access Control: The Cellular device may have files that need to have restricted access to them. The device might access a database where some sort of role based access control is necessary.

– Location Detection: The actual location of a cellular device needs to be kept hidden for reasons of privacy of the user. With the move to IP-based networks, the issue arises that a user may be associated with an access point and therefore their location might be compromised.

- Viruses and Malware: With increased functionality provided in cellular systems, problems prevalent in larger systems such as viruses and malware arise. The first virus that appeared on cellular devices was Liberty. An affected device can also be used to attack the cellular network infrastructure by becoming part of a large-scale denial of service attack.

- Device Security: If a device is lost or stolen, it needs to be protected from unauthorized use so that potential sensitive information such as e-mails, documents, phone numbers, etc., cannot be accessed.

Because of the massive architecture of a cellular network, there are a variety of attacks that the infrastructure is open to.

- Denial of Service (DOS): This is probably the most potent attack that can bring down the entire network infrastructure. This is caused by sending excessive data to the network, more than the network can handle, resulting in users being unable to access network resources.

- Distributed Denial of Service (DDOS): It might be difficult to launch a large-scale DOS attack from a single host. A number of hosts can be used to launch an attack.

- Channel Jamming: Channel jamming is a technique used by attackers to jam the wireless channel and therefore deny access to any legitimate users in the network.

- Unauthorized Access: If a proper method of authentication is not deployed, then an attacker can gain free access to a network and then can use it for services that he might not be authorized for.

- Eavesdropping: If the traffic on the wireless link is not encrypted then an attacker can eavesdrop and intercept sensitive communication such as confidential calls, sensitive documents, etc.

- Message Forgery: If the communication channel is not secure, then an attacker can intercept messages in both directions and change the content without the users ever knowing.

- Message Replay: Even if communication channel is secure, an attacker can intercept an encrypted message and then replay it back at a later time and the user might not know that the packet received is not the right one.

- Man In The Middle Attack: An attacker can sit in between a cell phone and an access station and intercept messages in between them and change them.

     – Session Hijacking: A malicious user can highjack an already estab-
       lished session and can act as a legitimate base station.

The cellular network security architecture provides features such as authentication, confidentiality, integrity, etc. Also, the WAP protocol
makes use of network security layers such as TLS/WTLS/SSL to provide a secure path for HTTP communication. Although 3G provides
good security features, there are always new security issues that come
up and researchers are actively pursuing new and improved solutions for
these issues. People have also started looking ahead at how new features
of the 4G network infrastructure will affect security and what measures
can be taken to add new security features and also improve upon those
that have been employed in 3G.

- Identify Management:

Subscriber access in GSM networks is based on a hardware token SIM
(Subscriber Identity Module) that is placed into any cellular mobile
phone. Each SIM contains a secret unique symmetric key (specified as
ki) stored together with the ID of the subscriber. This key is only shared
with the authentication center (AuC) of that GSM network operator
that issued the SIM.

When a GSM subscriber tries to log on to the GSM network (usually
when he switches on the phone) the SIM passes the subscriber's ID
to the AuC. The AuC then checks whether the SIM also "knows" the
respective ki (a secret unique symmetric key): The AuC sends a random
challenge message to the subscriber's phone. The SIM in the phone has
to encrypt that challenge message with the ki and send it back to the
AuC. The AuC encrypts the same message with the local copy of the
ki and compares the results. If they match, the subscriber is granted
access to the GSM network.

As most mobile cellular network operators have roaming agreements
with partners all over the world, the SIM is a globally accepted credential
for mobile communication access based on a globally standardized and
interoperable identification and authentication infrastructure.

The SIM can also be an anonymous or pseudonymous credential, as
there is no technical requirement to combine its information with any
information on the person using the SIM. Obviously, for most postpaid
contracts the providers require a billing address, but for prepaid contracts this billing address is not needed.

The almost global dominance of the cellular network standard for mobile
communications has inspired the SIM as platforms for identity management and related applications.

     – Identity management can be integrated into the SIM-Hardware.

- Identity management can use GSM subscriber information as issued with the SIM.

- Identity management can use GSM subscriber information stored in the GSM network.

The first two approaches aim at supporting the ID management that already exists in applications by using the cellular network infrastructure. The third approach expands the cellular network ID and user management itself and allows new revenue models in mobile communications.

5.2.4 Wireless Ad-Hoc Networks

A Wireless Ad-hoc Network (WANET) is a decentralized wireless network. Wireless multi-hop ad-hoc networks are formed by a group of mobile users or mobile devices spread over a certain geographical area. We call the users or devices forming the network nodes. The service area of the ad-hoc network is the whole geographical area where nodes are distributed. Each node is equipped with a radio transmitter and receiver that allows it to communicate with the other nodes. As mobile ad-hoc networks are self-organized networks, communication in ad-hoc networks does not require a central base station. Each node of an ad-hoc network can generate data for any other node in the network. All nodes can function, if needed, as relay stations for data packets to be routed to their final destination. A mobile ad-hoc network may be connected through dedicated gateways, or nodes functioning as gateways, to other fixed networks or the Internet. In this case, the mobile ad-hoc network expands the access to fixed network services.

The earliest wireless ad-hoc networks were the "packet radio" networks (PRNETs) from the 1970s, sponsored by DARPA after the ALOHAnet project.

The decentralized nature of wireless ad-hoc networks makes them suitable for a variety of applications where central nodes can't be relied on, and may improve the scalability of wireless ad-hoc networks compared to wireless managed networks, though theoretical and practical limits to the overall capacity of such networks have been identified.

Minimal configuration and quick deployment make ad-hoc networks suitable for emergency situations like natural disasters or military conflicts. The presence of a dynamic and adaptive routing protocol will enable ad-hoc networks to be formed quickly.

Wireless ad-hoc networks have certain advantages over the traditional communication networks. Some of these advantages are:

- Use of ad-hoc networks can increase mobility and flexibility, as ad-hoc networks can be brought up and torn down in a very short time.

- Ad-hoc networks can be more economical in some cases, as they eliminate fixed infrastructure costs and reduce power consumption at mobile nodes.

- Ad-hoc networks can be more robust than conventional wireless networks because of their nonhierarchical distributed control and management mechanisms.

- Because of multi-hop support in ad-hoc networks, communication beyond the Line of Sight (LOS) is possible at high frequencies.

- Multi-hop ad-hoc networks can reduce the power consumption of wireless devices. More transmission power is required for sending a signal over any distance in one long hop than in multiple shorter hops. It can easily be proved that the gain in transmission power consumption is proportional to the number of hops made.

- Because of short communication links (multi-hop node-to-node communication instead of long-distance node to central base station communication), radio emission levels can be kept low. This reduces interference levels, increases spectrum reuse efficiency, and makes it possible to use unlicensed unregulated frequency bands.

Typical applications of wireless ad-hoc networks:

Typical applications of ad-hoc networks are as follows [Dha07]:

- Mobile conferencing: Ad-hoc networks enable mobile conferencing for business users who need to collaborate outside their office where no network infrastructure is available. There is a growing need for mobile computing environment where different members of a project need to collaborate on design and development. The users need to share documents, upload, and download files and exchange ideas.

- Personal area and home networking: Ad-hoc networks are quite suitable for home as well as personal area networking applications. Mobile devices with Bluetooth or WLAN cards can be easily configured to form an ad-hoc network. With the Internet connectivity at home, these devices can easily be connected to Internet. Hence, the use of these kinds of ad-hoc networks has practical applications and usability.

- Emergency services: When the existing network infrastructure ceased to operate or damaged due to some kind of disaster like earthquakes, hurricanes, fire, and so on, ad-hoc networks can be easily deployed to provide solutions to emergency services. These networks can also be used for search and rescue operations, retrieval of patient data remotely from hospital, and many other useful services.

- Public hotspots: In places like airports, train stations, coffee shops, and pubs, football ground, malls, ad-hoc networks provide users to create their own network and communicate with each other instantly. Ad-hoc networks can also be used for entertainment purposes like providing instant connectivity for multi-user games. In addition, household Internet connectivity can be provided by a community hotspot.

- Military applications: In battlefield, MANET can be deployed for communications among the soldiers in the field. Different military units are expected to communicate and to cooperate with each other and within a specified area. In these kinds of low mobility environments, MANET is used for communications where virtually no network infrastructure is available. For example, mesh network is an ad-hoc peer-to-peer multi-hop network with no infrastructure. The important features are it is low in cost, with nodes being mobile, self-organized, self-balancing, and self-healing. It is easy to scale.

- Mobile commerce: Ad-hoc networks can be used to make electronic payments anytime, anywhere. Business users can retrieve customer/sales related information dynamically and can build reports on the fly.

- Ubiquitous and embedded computing applications: With the emergence of new generations of intelligent portable mobile devices, ubiquitous computing is becoming a reality. Ubiquitous computers will be around us, always doing some tasks for us without our conscious effort. These machines will also react to changing environment and work accordingly. These mobile devices will form an ad-hoc network and, gather various localized information and sometimes inform the users automatically.

- Location-based services: MANET when integrated with location-based information provides useful services. GPS (Global Positioning System), a satellite-based radio navigation system, is a very effective tool to determine the physical location of a device. A mobile host in a MANET when connected to a GPS receiver will be able to determine its current physical location. Another good example will be a group of tourists using PDAs with wireless LAN cards installed in them along with GPS connectivity. These mobile devices can be configured to form a MANET. These tourists can then exchange messages and locate each other using this MANET. Again consider the following scenario. Vehicles on highway can form an ad-hoc network to exchange traffic information. In addition, location-based information services can be delivered by MANETs. For example, one can advertise location-specific information like restaurants, shopping mall (push) and retrieve location-dependant information like travel guide, movie theater, drug store, and so forth (pull).

Wireless ad-hoc networks can be further classified as:

Wireless Personal Area Network (WPAN)

A personal area network (PAN) is a computer network used for communication among computer devices (including telephones and personal digital assistants) close to one person. The devices may or may not belong to the person in question. The reach of a PAN is typically a few meters. PANs can be used for communication among the personal devices themselves (intrapersonal communication), or for connecting to a higher-level network and the Internet (an uplink). Personal area networks may be wired with computer buses such as USB and FireWire. A wireless personal area network (WPAN) can also be made possible with network technologies such as IrDA, Bluetooth, UWB, Z-Wave, and ZigBee.

- **The Infrared Data Association (IrDA)** defines physical specifications communications protocol standards for the short-range exchange of data over infrared light, for uses such as personal area networks (PANs). IrDA is a very short-range example of free space optical communication. IrDA interfaces are used in palmtop computers, mobile phones, and laptop computers (most laptops and phones also offer Bluetooth, but it is now becoming more common for Bluetooth to simply replace IrDA in new versions of products). IrDA specifications include IrPHY (Infrared Physical Layer Specification), IrLAP (Infrared Link Access Protocol), IrLMP (Infrared Link Management Protocol), IrCOMM (Infrared Communications Protocol), Tiny TP (Tiny Transport Protocol), IrOBEX (Infrared Object Exchange), IrLAN (Infrared Local Area Network), and IrSimple. IrDA has now produced another standard, IrFM, for Infrared financial messaging (i.e., for making payments) also known as "Point & Pay." For the devices to communicate via IrDA, they must have a direct line of sight.

- **Bluetooth** is a wireless protocol for exchanging data over short distances from fixed and mobile devices, creating personal area networks (PANs). It was originally conceived as a wireless alternative to RS232 data cables. It can connect several devices, overcoming problems of synchronization.

  Bluetooth uses a radio technology called frequency-hopping spread spectrum, which chops up the data being sent and transmits chunks of it on up to 79 frequencies. In its basic mode, the modulation is Gaussian frequency-shift keying (GFSK). It can achieve a gross data rate of 1 Mb/s. Bluetooth provides a way to connect and exchange information between devices such as mobile phones, telephones, laptops, personal computers, printers, Global Positioning System (GPS) receivers, digital cameras, and video game consoles through a secure, globally unlicensed Industrial, Scientific, and Medical (ISM) 2.4 GHz short-range radio frequency bandwidth. The Bluetooth specifications are developed and licensed by the Bluetooth Special Interest Group (SIG). The Bluetooth

SIG consists of companies in the areas of telecommunication, computing, networking, and consumer electronics.

The features of the Bluetooth technology is as following:

- It separates the frequency band into hops. This spread spectrum is used to hop from one channel to another, which adds a strong layer of security.

- Up to eight devices can be networked in a piconet (the Bluetooth and 802.15 designation for a special personal area network (PAN)).

- Signals can be transmitted through walls and briefcases, thus eliminating the need for line-of-sight.

- Devices do not need to be pointed at each other, as signals are omni-directional.

- Both synchronous and asynchronous applications are supported, making it easy to implement on a variety of devices and for a variety of services, such as voice and Internet.

- Governments worldwide regulate it, so it is possible to utilize the same standard wherever one travels.

- **Near Field Communication (NFC)**, is a short-range high-frequency wireless communication technology which enables the exchange of data between devices over about a 10 centimeter (around 4 inches) distance. The technology is a simple extension of the ISO 14443 proximity-card standard (contactless card, RFID) that combines the interface of a smartcard and a reader into a single device. An NFC device can communicate with both existing ISO 14443 smartcards and readers, as well as with other NFC devices, and is thereby compatible with existing contactless infrastructure already in use for public transportation and payment. NFC is primarily aimed at usage in mobile phones.

  Like ISO 14443, NFC communicates via magnetic field induction, where two loop antennas are located within each other's near field, effectively forming an air-core transformer. It operates within the globally available and unlicensed radio frequency ISM band of 13.56 MHz, with a bandwidth of almost 2 MHz. It's Working distance with compact standard antennas: up to 20 cm. It can support data rates: 106kbit/s, 212kbit/s, or 424 kbit/s. There are two modes:

  Passive Communication Mode: The Initiator device provides a carrier field and the target device answers by modulating existing field. In this mode, the Target device may draw its operating power from the Initiator-provided electromagnetic field, thus making the Target device a transponder.

  Active Communication Mode: Both Initiator and Target device communicate by alternately generating their own field. A device deactivates its

RF field while it is waiting for data. In this mode, both devices typically need to have a power supply.

NFC technology is currently mainly aimed at being used with mobile phones. There are three main use cases for NFC:

- card emulation: the NFC device behaves like an existing contactless card
- reader mode: the NFC device is active and read a passive RFID tag, for example, for interactive advertising
- P2P mode: two NFC devices are communicating together and exchanging information.

NFC and Bluetooth are both short-range communication technologies, which have recently been integrated into mobile phones. The significant advantage of NFC over Bluetooth is the shorter set-up time. Instead of performing manual configurations to identify Bluetooth devices, the connection between two NFC devices is established at once (under a tenth of a second). To avoid the complicated configuration process, NFC can be used for the set-up of wireless technologies, such as Bluetooth. The maximum data transfer rate of NFC (424 kbit/s) is slower than Bluetooth (2.1 mbit/s). With less than 20 cm, NFC has a shorter range, which provides a degree of security and makes NFC suitable for crowded areas where correlating a signal with its transmitting physical device (and by extension, its user) might otherwise prove impossible. In contrast to Bluetooth, NFC is compatible with existing RFID structures. NFC can also work when one of the devices is not powered by a battery (e.g., on a phone that may be turned off, a contactless smart credit card, a smart poster).

- **Ultra-wide Band (UWB)** is a radio technology that can be used at very low energy levels for short-range high-bandwidth communications by using a large portion of the radio spectrum. This method is using pulse-coded information with sharp carrier pulses at a bunch of center frequencies in logical connex. UWB has traditional applications in non cooperative radar imaging. Most recent applications target sensor data collection, precision locating, and tracking applications.

UWB communications transmit in a way that doesn't interfere largely with other more traditional "narrow band" and continuous carrier wave uses in the same frequency band. However, first studies show that the rise of noise level by a number of UWB transmitters puts a burden on existing communications services. This may be hard to bear for traditional systems designs and may affect the stability of such existing systems.

UWB for transmitting information spread over a large bandwidth (> 500 MHz) that should, in theory and under the right circumstances, be able to share spectrum with other users. Regulatory settings of FCC are intended to provide an efficient use of scarce radio bandwidth while enabling both high data rate personal-area network (PAN) wireless connectivity, and longer-range, low data rate applications as well as radar and imaging systems.

UWB was traditionally accepted as pulse radio, but the FCC and ITU-R now define UWB in terms of a transmission from an antenna for which the emitted signal bandwidth exceeds the lesser of 500 MHz or 20% of the center frequency. Thus, pulse-based systems wherein each transmitted pulse instantaneously occupies the UWB bandwidth, or an aggregation of at least 500 MHz worth of narrow band carriers, for example, in orthogonal frequency-division multiplexing (OFDM) fashion can gain access to the UWB spectrum under the rules. Pulse repetition rates may be either low or very high. Pulse-based radars and imaging systems tend to use low repetition rates, typically in the range of 1 to 100 megapulses per second. On the other hand, communications systems favor high repetition rates, typically in the range of 1 to 2 giga-pulses per second, thus enabling short-range gigabit-per-second communications systems. Each pulse in a pulse-based UWB system occupies the entire UWB bandwidth, thus reaping the benefits of relative immunity to multipath fading (but not to intersymbol interference), unlike carrier-based systems that are subject to both deep fades and intersymbol interference.

Due to the extremely low emission levels currently allowed by regulatory agencies, UWB systems tend to be short-range and indoors applications. However, due to the short duration of the UWB pulses, it is easier to engineer extremely high data rates, and data rate can be readily traded for range by simply aggregating pulse energy per data bit using either simple integration or by coding techniques. Conventional OFDM technology can also be used subject to the minimum bandwidth requirement of the regulations. High data rate UWB can enable wireless monitors, the efficient transfer of data from digital camcorders, wireless printing of digital pictures from a camera without the need for an intervening personal computer, and the transfer of files among cell phone handsets and other handheld devices like personal digital audio and video players.

UWB is used as a part of location systems and real-time location systems. The precision capabilities combined with the very low power makes it ideal for certain radio frequency sensitive environments such as hospitals and healthcare. Another benefit of UWB is the short broadcast time which enables implementers of the technology to install orders of magnitude more transmitter tags in an environment relative to competitive technologies.

UWB is also used in "see-through-the-wall" precision radar imaging technology, precision locating and tracking (using distance measurements between radios), and precision time-of-arrival-based localization approaches. It exhibits excellent efficiency with a spatial capacity of approximately 1013 bit/s/m^2.

UWB has been a proposed technology for use in personal area networks and appeared in the IEEE 802.15.3a draft PAN standard. However, after several years of deadlock, the IEEE 802.15.3a task group has been dissolved in 2006.

- **Wireless USB (WUSB: 802.15.3)** is a new short-ranged, high-bandwidth wireless extension to USB intended to combine the speed and security of wired technology with the ease-of-use of wireless technology. WUSB is based on Ultra-WideBand (UWB) wireless technology defined by WiMedia Alliance, capable of sending 480 Mbit/s at distances up to 3 meters, and 110 Mbit/s at up to 10 meters. It operates in the 3.1 – 10.6 GHz band-range and spreads communication over an ultra-wideband of frequencies.

  IEEE 802.15.3a was an attempt to provide a higher-speed ultra-wideband physical layer (PHY) enhancement amendment to IEEE 802.15.3 for applications which involve imaging and multimedia. IEEE 802.15.3a UWB standardization attempt failed due to contrast between WiMedia Alliance and UWB Forum. On January 19, 2006, IEEE 802.15.3a task group (TG3a) members voted to withdraw the December 2002 project authorization request (PAR) that initiated the development of high data rate UWB standards. The IEEE 802.15.3a most commendable achievement was the consolidation of 23 UWB PHY specifications into two proposals using: Multi-Band Orthogonal frequency-division multiplexing (MB-OFDM) UWB, supported by the WiMedia Alliance, and Direct Sequence -UWB (DS-UWB), supported by the UWB Forum.

  Wireless USB is used in game controllers, printers, scanners, digital cameras, MP3 players, hard disks, and flash drives. It is also suitable for transferring parallel video streams.

- **Z-Wave** is a wireless communications standard designed for home automation, specifically to remote control applications in residential and light commercial environments. The technology, which is developed by Danish company Zensys, uses a low-power RF radio embedded or retrofitted into home electronics devices and systems, such as lighting, home access control, entertainment systems, and household appliances. The technology has been standardized by the Z-Wave Alliance, an international consortium of manufacturers that oversees interoperability between Z-Wave products and enabled devices.

  Z-Wave is a low-power wireless technology designed specifically for remote control applications. Unlike Wi-Fi and other IEEE 802.11-based wireless LAN systems that are designed primarily for high-bandwidth data flow, the Z-Wave RF system operates in the sub-Gigahertz frequency range and is optimized for low-overhead commands such as on-off (as in a light switch or an appliance) and raise-lower (as in a thermostat or volume control), with the ability to include device metadata in the communications. Because Z-Wave operates apart from the 2.4 GHz frequency of 802.11 based wireless systems, it is largely impervious to interference from common household wireless electronics, such as Wi-Fi routers, cordless telephones, and Bluetooth devices that work in the same frequency range. This freedom from household interference allows for a standardized low-bandwidth control medium that can be reliable alongside common wireless devices.

As a result of its low power consumption and low cost of manufacture, Z-Wave is easily embedded in consumer electronics products, including battery-operated devices such as remote controls, smoke alarms, and security sensors. Z-Wave is currently supported by over 200 manufacturers worldwide and appears in a broad range of consumer products in the U.S. and Europe.

The standard itself is not open and is available only to Zensys customers under non-disclosure agreement. Some Z-Wave product vendors have embraced the open source and hobbyist communities.

Z-Wave is a mesh networking technology where each node or device on the network is capable of sending and receiving control commands through walls or floors and around household obstacles or radio dead spots that might occur in the home. Z-Wave devices can work singly or in groups, and can be programmed into scenes or events that trigger multiple devices, either automatically or via remote control. Some common applications for Z-Wave include: remote home control and management, energy conservation, home safety and security systems, and home entertainment.

- **ZigBee** is the name of a specification for a suite of high-level communication protocols using small, low-power digital radios based on the IEEE 802.15.4-2006 standard for wireless personal area networks (WPANs), such as wireless headphones connecting with cell phones via short-range radio. The technology is intended to be simpler and less expensive than other WPANs, such as Bluetooth. ZigBee is targeted at radio-frequency (RF) applications that require a low data rate, long battery life, and secure networking.

The ZigBee Alliance is a group of companies that maintain and publish the ZigBee standard.

ZigBee is a low-cost, low-power, wireless mesh networking standard. The low cost allows the technology to be widely deployed in wireless control and monitoring applications, the low power-usage allows longer life with smaller batteries, and the mesh networking provides high reliability and larger range.

ZigBee protocols are intended for use in embedded applications requiring low data rates and low power consumption. ZigBee's current focus is to define a general-purpose, inexpensive, self-organizing mesh network that can be used for industrial control, embedded sensing, medical data collection, smoke and intruder warning, building automation, home automation, etc. The resulting network will use very small amounts of power, individual devices must have a battery life of at least two years to pass ZigBee certification.

Typical application areas include: home entertainment and control, home awareness, mobile services, commercial building and industrial plant.

Wireless Local Area Network (WLAN)

A wireless LAN or WLAN or wireless local area network is the linking of two or more computers or devices using spread-spectrum or OFDM modulation technology based to enable communication between devices in a limited area. This gives users the mobility to move around within a broad coverage area and still be connected to the network.

The popularity of wireless LANs is a testament primarily to their convenience, cost efficiency, and ease of integration with other networks and network components. The majority of computers sold to consumers today come pre-equipped with all necessary wireless LAN technology. Benefits of wireless LANs include:

- Convenience: The wireless nature of such networks allows users to access network resources from nearly any convenient location within their primary networking environment (home or office). With the increasing saturation of laptop-style computers, this is particularly relevant.

- Mobility: With the emergence of public wireless networks, users can access the internet even outside their normal work environment. Most chain coffee shops, for example, offer their customers a wireless connection to the internet at little or no cost.

- Productivity: Users connected to a wireless network can maintain a nearly constant affiliation with their desired network as they move from place to place. For a business, this implies that an employee can potentially be more productive as his or her work can be accomplished from any convenient location. For example, a hospital or warehouse may implement Voice over WLAN applications that enable mobility and cost savings.

- Deployment: Initial setup of an infrastructure-based wireless network requires little more than a single access point. Wired networks, on the other hand, have the additional cost and complexity of actual physical cables being run to numerous locations (which can even be impossible for hard-to-reach locations within a building).

- Expandability: Wireless networks can serve a suddenly increased number of clients with the existing equipment. In a wired network, additional clients would require additional wiring.

- Cost: Wireless networking hardware is at worst a modest increase from wired counterparts. This potentially increased cost is almost always more than outweighed by the savings in cost and labor associated to running physical cables.

Wireless Metropolitan Area Network (WMAN)

Metropolitan area networks, or MANs, are large computer networks usually spanning a city.

A MAN is optimized for a larger geographical area than a LAN, ranging from several blocks of buildings to entire cities. MANs can also depend on communications channels of moderate-to-high data rates. A MAN might be owned and operated by a single organization, but it usually will be used by many individuals and organizations. MANs might also be owned and operated as public utilities. They will often provide means for internetworking of local networks. Metropolitan area networks can span up to 50 km, devices used are modem and wire/cable. WMANs typically use wireless infrastructure or Optical fiber connections to link their sites.

WiMAX (Worldwide Interoperability for Microwave Access) is the term used to refer to wireless MANs and is covered in IEEE 802.16d/802.16e. WiMAX is a telecommunications technology that provides wireless transmission of data using a variety of transmission modes, from point-to-point links to portable internet access. The technology provides up to 75 Mb/s symmetric broadband speed without the need for cables. WiMAX is "a standards-based technology enabling the delivery of last mile wireless broadband access as an alternative to cable and DSL."

WPAN, WLAN, and WMAN networks were designed for portable terminals, often in a single-cell configuration. They cover specifications for the physical layer and the data link layer of the OSI model. These systems can handle mobile stations but with serious restrictions.

Mobile Ad-Hoc Networks (MANETs)

A mobile ad-hoc network (MANET) is a kind of wireless ad-hoc network, and is a self-configuring network of mobile routers (and associated hosts) connected by wireless links, the union of which form an arbitrary topology.

The routers are free to move randomly and organize themselves arbitrarily; thus, the network's wireless topology may change rapidly and unpredictably. Such a network may operate in a stand-alone fashion, or may be connected to the larger Internet. Minimal configuration and quick deployment make ad-hoc network suitable for emergency situations like natural or human-induced disasters, military conflicts, emergency medical situations, etc.

The popular IEEE 802.11 wireless protocol incorporates an ad-hoc networking system when no wireless access points are present, although it would be considered a very low-grade ad-hoc protocol by specialists in the field. The IEEE 802.11 system only handles traffic within a local "cloud" of wireless devices. Each node transmits and receives data but does not route anything between the network's systems. However, higher-level protocols can be used to aggregate various IEEE 802.11 ad-hoc networks into MANETs.

Approaches are intended to be relatively lightweight in nature, suitable for multiple hardware and wireless environments, and address scenarios where MANETs are deployed at the edges of an IP infrastructure. Hybrid mesh infrastructures (e.g., a mixture of fixed and mobile routers) should also be supported by MANET specifications and management features.

Types of MANET:

- Vehicular Ad-Hoc Networks (VANET) are a form of MANETs used for communication among vehicles and between vehicles and roadside equipment.

- Intelligent vehicular ad-hoc network (InVANET) is a kind of Intelligence in Vehicle(s), which provide multiple autonomic intelligent solutions to make automotive vehicles to behave in intelligent manner during vehicle-to-vehicle collisions, accidents, drunken driving, etc.

Wireless Mesh Networks

A wireless mesh network (WMN) is a communications network made up of radio nodes organized in a mesh topology. The coverage area of the radio nodes working as a single network is sometimes called a mesh cloud. Access to this mesh cloud is dependent on the radio nodes working in harmony with each other to create a radio network. A mesh network is reliable and offers redundancy. When one node can no longer operate, the rest of the nodes can still communicate with each other, directly or through one or more intermediate nodes. The animation below illustrates how wireless mesh networks can self form and self heal.

A wireless mesh network can be seen as a type of wireless ad-hoc network, where all radio nodes are static and doesn't experience direct mobility.

Wireless mesh architecture is a first step toward providing high-bandwidth network over a specific coverage area. Wireless mesh architecture's infrastructure is, in effect, a router network minus the cabling between nodes. It's built of peer radio devices that don't have to be cabled to a wired port like traditional WLAN access points (AP) do. Mesh architecture sustains signal strength by breaking long distances into a series of shorter hops. Intermediate nodes not only boost the signal, but cooperatively make forwarding decisions based on their knowledge of the network, i.e., performs routing. Such an architecture may with careful design provide high bandwidth, spectral efficiency, and economic advantage over the coverage area.

Example of three types of wireless mesh network:

- Infrastructure wireless mesh networks: Mesh routers form an infrastructure for clients.

- Client wireless mesh networks: Client nodes constitute the actual network to perform routing and configuration functionalities.

- Hybrid wireless mesh networks: Mesh clients can perform mesh functions with other mesh clients as well as accessing the network.

Wireless mesh network have a relatively stable topology except for the occasional failure of nodes or addition of new nodes. The traffic, being aggregated from a large number of end users, changes infrequently. Practically all the traffic in an infrastructure mesh network is either forwarded to or from a gateway, while in ad-hoc networks or client mesh networks the traffic flows between arbitrary pairs of nodes.

The infrastructure of WMNs can be decentralized (with no central server) or centrally managed (with a central server), both are relatively inexpensive, and very reliable and resilient, as each node needs only transmit as far as the next node. Nodes act as routers to transmit data from nearby nodes to peers that are too far away to reach in a single hop, resulting in a network that can span larger distances. The topology of a mesh network is also more reliable, as each node is connected to several other nodes. If one node drops out of the network, due to hardware failure or any other reason, its neighbors can find another route using a routing protocol.

Mesh networks may involve either fixed or mobile devices (see Figure 5.11). The solutions are as diverse as communication needs, for example in difficult environments such as emergency situations, tunnels, and oil rigs to battlefield surveillance and high-speed mobile video applications on board public transport or real-time racing car telemetry. A significant application for wireless mesh networks is VoIP. By using a Quality of Service scheme, the wireless mesh may support local telephone calls to be routed through the mesh.

Figure 5.12 shows an example for detailed connectivity of a backbone mesh network to WiFi, WiMAX, and wireless cellular networks. A WiFi network, a cellular network, and a WiMAX network can connect through mesh routers with gateway bridges. A router with gateway bridge capability enables the integration of WMNs with other type networks, although traditional routers with regular network interface cards (NICs) can connect to mesh networks.

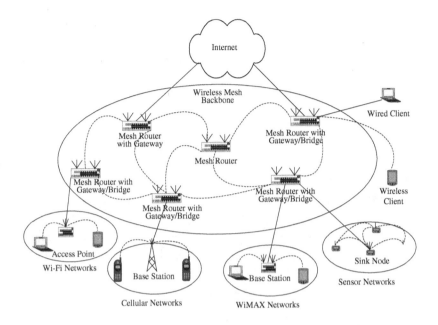

Figure 5.11: Architecture of Wireless Mesh Networks

Mesh users can also operate as routers for mesh networking, making the connectivity much simpler and faster than conventional wireless networks with base stations. Figure 5.13 shows another scenario, in which a wireless mesh network backbone is connected to wireless mesh users. Users are communicating in an ad-hoc fashion, with each individual user acting as a router and connected to a mesh router gateway. In this network, wired users, as shown by a LAN in the figure, can also be connected to WMN, using a mesh router gateway. User meshing provides peer-to-peer networks among users.

The inclusion of multiple wireless interfaces in mesh routers significantly enhances the flexibility of mesh networks. WMNs offer advantages of low-cost, easy network maintenance, and remarkably more reliable service coverage than conventional ad-hoc networks. Mesh networks are being designed for metropolitan and enterprise networking, and most of standards, such as IEEE 802.11, IEEE 802.15, and IEEE 802.16, are accepted in WMN infrastructures. A widely accepted radio technology is the series of IEEE 802.11 standards.

The benefits of a mesh network are as follows:

- Scalability. The WMN infrastructure is designed to be scalable as the need for network access increases.

- Ad-hoc networking support. WMNs have the capability to self-organize and be connected to certain points of ad-hoc networks for a short period of time.

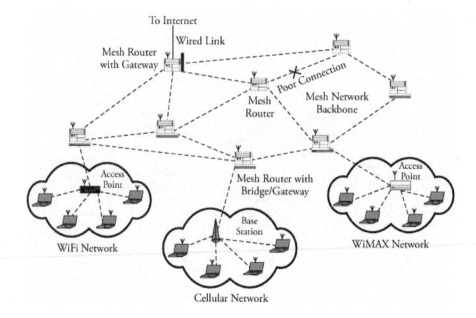

Figure 5.12: Overview of a Backbone Mesh Network and Connections to WiFi, WiMAX, and Wireless Cellular Networks

- Mobility support of end nodes. End nodes are supported through the wireless infrastructure.

- Connectivity to wired infrastructure. Gateway mesh routers may integrate heterogeneous networks in both wired and wireless fashions.

To achieve scalability in WMNs, all protocols from the MAC layer to the application layer must be scalable. "Topology-" and "routing-aware" MAC can substantially improve the performance of WMNs.

The QoS provisioning in wireless mesh networks is different from that of classical ad-hoc networks. Several applications are broadband services with heterogeneous QoS requirements. Consequently, additional performance metrics, such as delay jitter and aggregate and per node throughput, must be considered in establishing a route. Application-specific security protocols must also be designed for WMNs. Security protocols for ad-hoc networks cannot provide any reliability, as the traffic in such networks can resemble the one flowing in the wired Internet.

Wireless Sensor Networks

A wireless sensor network (WSN) is a wireless network consisting of spatially distributed autonomous devices using sensors to cooperatively monitor

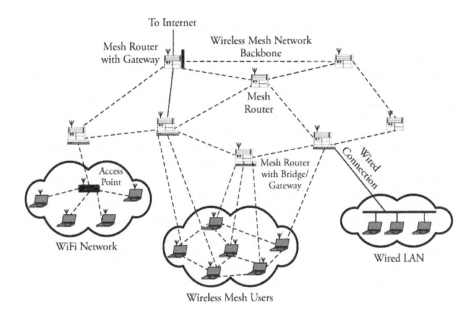

Figure 5.13: Connectivity between a Backbone Wireless Mesh Network to Wireless Users and Other Networking Devices

physical or environmental conditions, such as temperature, sound, vibration, pressure, motion or pollutants, at different locations. The development of wireless sensor networks was originally motivated by military applications such as battlefield surveillance. However, wireless sensor networks are now used in many civilian application areas, including environment and habitat monitoring, healthcare applications, home automation, and traffic control.

In addition to one or more sensors, each node in a sensor network is typically equipped with a radio transceiver or other wireless communications device, a small microcontroller, and an energy source, usually a battery. The envisaged size of a single sensor node can vary from shoebox-sized nodes down to devices the size of grain of dust, although functioning "motes" of genuine microscopic dimensions have yet to be created. The cost of sensor nodes is similarly variable, ranging from hundreds of dollars to a few cents, depending on the size of the sensor network and the complexity required of individual sensor nodes. Size and cost constraints on sensor nodes result in corresponding constraints on resources such as energy, memory, computational speed, and bandwidth.

A sensor network normally constitutes a wireless ad-hoc network, meaning that each sensor supports a multi-hop routing algorithm (several nodes may forward data packets to the base station).

Figure 5.14 shows the architecture for the applications of wireless sensor networks [Le05].

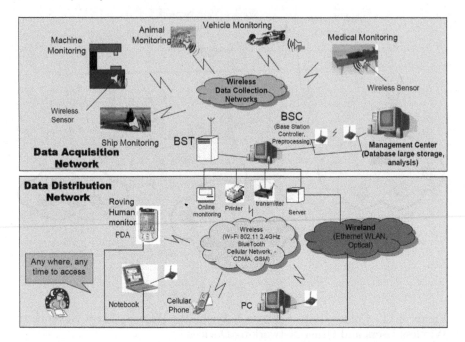

Figure 5.14:   Architecture for Applications of Wireless Sensor Networks

The applications for WSNs are many and varied, but typically involve some kind of monitoring, tracking, and controlling. Specific applications for WSNs include habitat monitoring, object tracking, nuclear reactor control, fire detection, and traffic monitoring. In a typical application, a WSN is scattered in a region where it is meant to collect data through its sensor nodes.

Unique characteristics of a WSN include:

- Limited power they can harvest or store

- Ability to withstand harsh environmental conditions

- Ability to cope with node failures

- Mobility of nodes

- Dynamic network topology

- Communication failures

- Heterogeneity of nodes

- Large scale of deployment

- Unattended operation

Sensor nodes can be imagined as small computers, extremely basic in terms of their interfaces and their components. They usually consist of a processing unit with limited computational power and limited memory, sensors (including specific conditioning circuitry), a communication device (usually radio transceivers or alternatively optical), and a power source usually in the form of a battery. Other possible inclusions are energy harvesting modules, secondary ASICs, and possibly secondary communication devices (e.g., RS-232 or USB).

The base stations are one or more distinguished components of the WSN with much more computational, energy, and communication resources. They act as a gateway between sensor nodes and the end user.

5.2.5 Management of Wireless Ad-Hoc Networks

Wireless ad-hoc network management refers to use of software tools designed to enhance performance, reliability, and security of wireless networks particularly by diagnosing, detecting and reducing sources of radio frequency interference. Unlike in the case of wired networks, where the redundancy of the equipment can be applied to mitigate the impact of performance problems and network failures to some extent, the wireless networking have very limited options owing to availability of very limited wireless spectrum and the effects of wireless interference.

The factors that can have impact on the wireless network performance include traffic flows, working of the network topologies and network protocols, hardware, software, and the environmental conditions. Therefore, often the Wireless users can be subjected to problems such as lack of coverage, intermittent discontinuity, and difficult to monitor security aspects.

Wireless Network Management systems help to ensure network availability, as well as to provide other maintenance tasks, such as performance monitoring, testing, and fault management.

A fixed network management model is not directly appropriate for ad-hoc networks. New management paradigms and architectures must be defined to cope with the dynamics and resource constraints of such networks. Managing ad-hoc networks presents us with different research issues that are not encountered in the same way in the common fixed networks [BSF07] [MK04] [BSF07] [Dha07].

- **Mobility management**: The nodes may not be static in space and time resulting in a dynamic network topology. Nodes can move freely and independently. Also some new nodes can join the network and some nodes may leave the network. Individual random mobility, group mobility, updating along preplanned routes can have major impact on the selection of a routing scheme and can thus influence performance.

Multicasting becomes a difficult problem because mobility of nodes creates inefficient multicast trees and inaccurate configuration of network topology.

There are two distinct methods for mobility support in wireless ad-hoc networks: mobile IP, and fast-routing protocols. Mobile IP offers a pure network layer architectural solution for mobility support and isolates the higher layers from the impact of mobility. But mobile IP works not efficient on a large-scale network. Fast routing protocols are designed to cope with changes in the network topology. Routing in ad-hoc networks is basically a compromise between the method of dealing with fast topology changes and keeping the routing overhead minimal.

– Routing Management

Ad-hoc networks rely on a routing plane capable to fit with their dynamics and resource constraints. In particular, an ad-hoc routing protocol must be capable to maintain an updated view of the dynamic network topology while minimizing the routing traffic.

Dynamically changing network topology makes routing a complex issue and requires new approaches that should be considered in design of routing protocols for sensor networks.

A message from source node to destination node goes through multiple nodes because of limited transmission radius. Every node acts as a router and forwards packets from other nodes to facilitate multi-hop routing. Ad-hoc networks often exhibit multiple hops for obstacle negotiation, spectrum reuse, and energy conservation. Battlefield covert operations also favor a sequence of short hops to reduce detection by the enemy.

The table-driven protocols are the Destination-Sequenced Distance Vector (DSDV) protocol [PB94], the Cluster-Head Gateway Switch Routing (CGSR) protocol [LCW+97], and the Wireless Routing Protocol (WRP) [MG96].

The source-initiated protocols are the Dynamic Source Routing (DSR) protocol [RFC4728], the Associative-Based Routing (ABR) protocol [Toh97], Temporally Ordered Routing Algorithm (TORA) [WHB05], and Ad-Hoc On-Demand Distance Vector (AODV) protocol [RFC3561].

– Location Management:

Since there is no fixed infrastructure available for MANET with nodes being mobile in the three-dimensional space, location management becomes a very important issue. For example, route discovery and route maintenance are some of the challenges in designing routing protocols. In addition, finding the position of a node at a given time is an important problem. This led to the development of location-aware routing, which means that a node will be able to know its current position.

A useful application of location-based service for MANET will be in the area of navigation. When devices are equipped with wireless connectivity along with location-based information integrated with navigation systems, users can communicate with each other forming an ad-hoc network. Location-based emergency services are also potential applications of these systems. Another important application area is geocast, which means sending messages to all the hosts in a particular geographic region. Geocasting will be a very useful application when someone wants to send some messages to people in a particular region. This is particularly important in situations when there is a disaster or emergency.

There are quite a few protocols that have been proposed in the recent years specifically for location management. They can be classified into two major categories, location-assisted and zone-based protocols. Location-assisted protocols take advantage of local information of hosts. In zone-based routing, the network is divided into several nonoverlapping regions (zones) and each node belongs to a certain zone based on its physicals location.

- **Resource Management**:

  WANET uses wireless medium (radio, infrared, etc.) to transmit and receive data. Nodes can share the same media easily. WANET can be more economical in some cases, as they eliminate fixed infrastructure costs and reduce power consumption at mobile nodes.

  But WANET links have limited bandwidth and variable capacity. They are also error prone. In addition, most ad-hoc nodes (e.g., laptops, PDAs, sensors) have limited power supply and no capability to generate their own power (e.g., solar panels). Energy efficient protocol design (e.g., MAC, routing, resource discovery) is critical for longevity of the mission.

  Bandwidth and energy are scarce resources in ad-hoc networks. The wireless channel offers a limited bandwidth, which must be shared among the network nodes. The mobile devices are strongly dependent on the lifetime of their battery.

  - Spectrum management:

    The scarcity of spectrum gives rise to the need for technologies that can either squeeze more system capacity over a given band, or utilize the high bands that are typically less crowded. The challenge for the latter option is the development of equipment at the same cost and performance as equipment used for lower band systems. Because of short communication links (multi-hop node-to-node communication instead of long-distance node to central base station communication), radio emission levels can be kept low. This reduces interference levels, increases spectrum reuse efficiency, and makes it possible to use unlicensed unregulated frequency bands.

– Power management:

Portable hand-held devices have limited battery power. These devices often participate as nodes in a MANET and deliver and route packets. Whenever the power of a node is depleted, the WANET may cease to operate or may not function efficiently. To minimize energy consumption and prolong the life of the nodes in sensor networks, the design of sensor networks requires energy-awareness in each phase.

Power consumption of dynamic components is proportional to $CV^2 f$, where C is the capacitance of the circuit, V is the voltage swing and f is the clock frequency. Thus, to provide power efficiency, one or more of the following are needed: (a) greater levels of VLSI integration to reduce capacitance; (b) lower voltage circuits must be developed; (c) clock frequency must be reduced. Alternatively, wireless systems can be built in such a way that most of the processing is carried out in fixed parts of the network (such as Base Stations in cellular systems), which are not power-limited. Finally, the same concerns for power management also affect the software for wireless networks: efficient software can power down device parts when those are idle.

– Design of Energy-Efficient Protocols

In design of energy-efficient protocols, following aspects should be taken into account carefully:

1). End-to-end delay: Shortest path routing reduces delay, but may increase power consumption due to increased distance between two adjacent nodes in the path, since transmit power is proportional to some power of distance as mentioned in the previous section. Increased delay also means increased number of packets in the system and this could become a problem in terms of network capacity.

2). Packet loss rate: Low transmit power reduces power consumption but may increase the packet loss rate due to weaker signal strength. When packet loss rate increases, it results in retransmissions of the packets, again increasing total number of packets in the system.

3). Network capacity: Low transmit power also decreases the range of the transmitting node, which may result in less number of connections to neighboring nodes. Reduced number of links per node could also reduce the traffic carrying capacity of the network; worse, it can even break the network connectivity and leave the network into disjoint sub-networks.

4). Relaying overhead: Routing in MANET results in multi-hop transmissions because of low transmit power of the nodes require having more intermediate nodes from source to destination. Multi-hop routing adds relaying overhead to each node in the routing path, as each node in the path needs to receive the packet from physical layer, figure out the next hop and retransmit through the network stack. But multi-hop routing in some cases consumes less power than single-hop routing.

5). Interference: High power transmission also induces greater interference to neighboring nodes. Interference also results in unnecessary power consumption in the neighboring nodes, as they have to receive the signal even when the signal is not destined for them. Stronger interferences could cause increased number of collisions, thus increasing the power consumption even more.

6). Battery life: In minimum transmitted power routing, several other nodes for routing packets can use the same route. When this keeps happening over a period of time, the battery power on those nodes that are on the routing path will run out, thus reducing the network lifetime.

7). Connectivity: Sleep mode reduces power consumption by an order of magnitude. But when a number of nodes go into the sleep mode to save power, without coordination, it may disrupt network connectivity.

- **Security Management**:

WANETs are more vulnerable than fixed networks because of the nature of the wireless medium and the lack of central coordination. Wireless transmissions can be easily captured by an intruding node. A misbehaving node can perform a denial of service attack by consuming the bandwidth resources and making them unavailable to the other network nodes. It is also not easy to detect a malicious node in a multi-hop ad-hoc network and implement denial of service properly. In addition, in a multicasting scenario, traffic may pass through unprotected routers, which can easily get unauthorized access to sensitive information (as in the case with military applications). Dynamic topology and movement of nodes in an ad-hoc network make key management difficult if cryptography is used in the routing protocol.

Challenges of security management:

Types of attacks:

The ad-hoc networks, however, are even more vulnerable to attacks than the infrastructure counterparts. Both active and passive attacks are possible.

In a passive attack, the normal operation of a routing protocol is not interrupted. Instead, an intruder tries to gather information by listening. Active attacks can sometimes be detectable and thus are less important.

In an active attack, an attacker can insert some arbitrary packets of information into the network to disable it or to attract packets destined to other nodes. Typical attacks for WANET is as follows:

1) Pin Attack

With the pin, or black-hole, attack, a malicious node pretends to have the shortest path to the destination of a packet. Normally, the intruder listens to a path set-up phase and, when learns of a request for a route, sends a reply advertising a shortest route. Then, the intruder can be an official part of the network if the requesting node receives its malicious reply before the reply from a good node, and a forged route is set up. Once it becomes part of the network, the intruder can do anything within the network, such as undertaking a denial-of-service attack.

2) Location-Disclosure Attack

By learning the locations of intermediate nodes, an intruder can find out the location of a target node. The location-disclosure attack is made by an intruder to obtain information about the physical location of nodes in the network or the topology of the network.

3) Routing Table Overflow

Sometimes, an intruder can create routes whose destinations do not exist. This type of attack, known as the routing table overflow, overwhelms the usual flow of traffic, as it creates too many dummy active routes. This attack has a profound impact on proactive routing protocols, which discover routing information before it is needed, but minimal impact on reactive routing protocols, which create a route only when needed.

4) Energy-Exhaustion Attack

Battery-powered nodes can conserve their power by transmitting only when needed. But an intruder may try to forward unwanted packets or request repeatedly fake or unwanted destinations to use up the energy of nodes' batteries.

Security of IEEE 802.11

The security mechanisms for the wireless IEEE 802.11a and b standards known as wired equivalent privacy (WEP). WEP provides a level of security similar to that found in wired networks. It offers authentication and data encryption between a host and a wireless base station, using a secret shared key.

WEP is simple and relatively weak. A new protocol for IEEE 802.11i, called Extensible Authentication Protocol (EAP), specifies the interaction between a user and an authentication server. EAP security mechanism works as follows:

A base station first announces its presence and the types of security services it can provide to the wireless users. This way, users can request the appropriate type and level of encryption or authentication. EAP frames are encapsulated and sent over the wireless link. After decapsulation at the base station, the frames are encapsulated again, this time using a protocol called RADIUS for transmission over UDP to the authentication server. With EAP, public-key encryption is used, whereby a shared secret key known only to the user and the authentication server is created. The wireless user and the base station can also generate additional keys to perform the link-level encryption of data sent over the wireless link, which makes it much more secure than the one explained for 802.11a, b.

Security Routing Protocol

In order to prevent ad-hoc networks from attacks and vulnerability, a routing protocol must possess the following properties:

1) Authenticity. When a routing table is updated, it must check whether the updates were provided by authenticated nodes and users. The most challenging issue in ad-hoc networks is the lack of a centralized authority to issue and validate certificates of authenticity.

2) Integrity of information. When a routing table is updated, the information carried to the routing updates must be checked for eligibility. A misleading update may alter the flow of packets in the network.

3) In-order updates. Ad-hoc routing protocols must contain unique sequence numbers to maintain updates in order. Out-of-order updates may result in the propagation of wrong information.

4) Maximum update time. Updates in routing tables must be done in the shortest possible time to ensure the credibility of the update information. A timestamp or time-out mechanism can normally be a solution.

5) Authorization. An unforgeable credential along with the certificate authority issued to a node can determine all the privileges that the node can have.

6) Routing encryption. Encrypting packets can prevent unauthorized nodes from reading them, and only those routers having the decryption key can access messages.

7) Route discovery. It should always be possible to find any existing route between two points in a network.

8) Protocol immunization. A routing protocol should be immune to intruding nodes and be able identify them.

9) Node-privacy location. The routing protocol must protect the network from spreading the location or other unpublic information of individual nodes.

10) Self-stabilization. If the self-stabilization property of ad-hoc network performs efficiently, it must stabilize the network in the presence of damages continually received from malicious nodes.

11) Low computational load. Typically, an ad-hoc node is limited in powers as it uses a battery. As a result, a node should be given the minimal computational load to maintain enough power to avoid any denial-of-service attacks from low available power.

- **Self-management**: WANET has no centralized control which implies that network management will have to be distributed across various nodes. The dynamic nature of WANET requires that the management plane is able to adapt itself to the heterogeneous capacity of devices and to the environment modifications. The ad-hoc network is deployed in a spontaneous manner, the management architecture should be deployed in the same manner in order to minimize any human intervention. The complexity of mobile ad-hoc networks makes it imperative that any management system for such networks minimize the amount of human intervention required to achieve the desired network performance. This includes automatic fault detection and remediation as well as automation of component configuration. Self-management is appropriate to manage such dynamic and complex networks.

  In distributed and infrastructure-less environment, self-management enables the ad-hoc network must autonomously determine its own configuration parameters, including addressing, routing, clustering, position identification, neighborhood awareness, power control, etc. In some cases, special nodes (e.g., mobile backbone nodes) can coordinate their motion and dynamically distribute in the geographic area to provide coverage of disconnected islands.

- **Scalability Management**:

  In WANET, the management plane must be easy to maintain and must remain coherent, even when the ad-hoc network size is growing, when the network is merging with an other one or when it is splitting into different ones. Most of the routing algorithms are designed for relative small wireless ad-hoc networks. However, in some applications (e.g., large environmental sensor fabrics, battlefield deployments, urban vehicle grids, etc.), the ad-hoc network can grow to several thousand nodes. For example, there are some applications of sensor networks and tactical networks, which require deployment of large number of nodes. For wireless "infrastructure" networks, scalability is simply handled by a hierarchical construction. The limited mobility of infrastructure networks can also be easily handled using Mobile IP or local handoff techniques. In contrast, because of the more extensive mobility and the lack of fixed references, pure ad-hoc networks do not tolerate mobile IP or a fixed hierarchy structure. Thus, mobility, jointly with large scale is one of the most critical challenges in ad-hoc design.

- **Reliability Management**:

  Reliable data communications to a group of mobile nodes that continuously change their locations is extremely important particularly in emergency situations. Contrary to wireline, the wireless medium is highly unreliable. This is due to the variable capacity, limited bandwidth, limited battery power, attenuation and distortion in wireless signals, nodes motion and prone to error. The way wireless signals are distorted is difficult to predict, since distortions are generally of random nature. Thus, wireless systems must be designed with this fact in mind. Procedures for hiding the impairments of the wireless links from high-layer protocols and applications as well as development of models for predicting wireless channel behavior would be highly beneficial.

  Given the nature of WANETs, it is necessary to make the network management system survivable. This is due to the dynamic nature of such networks, as nodes can move in and out of the network and may be destroyed for various reasons such as battlefield casualties, low battery power, etc. Thus there should be no single point of failure for the network management system.

- **Integrated Management**: A typical ad-hoc network is composed of various devices such as laptops, personal digital assistants, phones, and intelligent sensors. These devices do not provide the same hardware and software capabilities, but they have to interoperate in order to establish a common network and implement a common task. Seamless roaming and handoff in heterogeneous networks, multiple radio integration and coordination, communication between wireless and wired networks and different administrative domains invoke WANET management systems to implement higher-level integrated and cooperated policies. The network maintenance tasks are shared among the mobile nodes in a distributed manner based on synchronization and cooperation mechanisms.

  Middleware is an important architectural system to support distributed applications. The role of middleware is to present a unified programming model to applications and to hide problems of heterogeneity and distribution. It provides a way to accommodate diverse strategies, offering access to a variety of systems and protocols at different levels of a protocol stack.

  Recent developments in the areas of wireless multimedia and mobility require more openness and adaptivity within middleware platforms. Programmable techniques offer a feasible approach to avoid timevarying QoS impairments in wireless and mobile networking environments. Multimedia applications require open interfaces to extend systems to accommodate new protocols. They also need adaptivity to deal with varying levels of QoS from the underlying network. Mobility aggravates these problems by changing the level of connectivity drastically over time.

- **Service Management**: Incorporating QoS in MANET is a non trivial problem because of the limited bandwidth and energy constraints. Designing protocols that support multi-class traffic and allows preemption, mobile nodes position identification, packet prioritization are some of the open areas of research. In order to provide end-to-end QoS guarantee, a coordinated effort is required for multi-layer integration of QoS provisioning. The success and future application of WANET will depend on how QoS will be guaranteed in the future.

Emerging multimedia services demand high bandwidth. There are some challenges and problems that need to be solved before real-time multimedia can be delivered over wireless links in ad-hoc networks. The need for more bandwidth, less delay and minimum packet loss are some of the criteria for high-quality transmission. However, the current best-effort network architecture does not offer any quality of service (QoS). Hence, in order to support multimedia traffic, efforts must be made to improve QoS parameters like end-to-end delay, packet loss ratio, and jitter.

5.3 Optical Networks

5.3.1 Introduction of Optical Networks

Optical networks are high-capacity telecommunications networks based on optical technologies and components that provide routing, grooming, and restoration at the wavelength level as well as wavelength-based services. Optical networks provide higher capacity and reduced costs for new applications such as the Internet, video and multimedia interaction, and advanced digital services.

According to the selection of the optical communications medium, optical networks can be divided into two categories, guided and unguided systems. In unguided systems the optical beam that is transmitted from the source widens as it propagates into space resembling microwave transmission. Guided systems use optical fiber as the communications medium and are thus also known as fiber-optics communication systems. Since the majority of today's optical communications systems are fiber-based, the term optical systems is often used as a synonym of guided systems despite the fact that there also exist unguided optical systems. Optical networks can be opaque or all-optical and can be of single-wavelength or based on wavelength division multiplexing (WDM).

Over years, developments in optical networking technology have driven new advances in network architectures, and the resulting methods in which

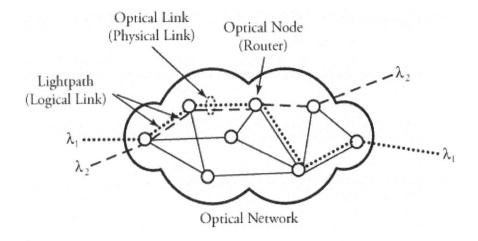

Figure 5.15: Overview of an Optical Network

those networks are managed. Optical networking technology advances have produced vast changes in network architectures for both public and private networks. Packet switching technology embedded in edge devices has been deployed at an increasing rate and is rapidly replacing circuit-switched hierarchical networks. These optical networks are designed with robust data communications capabilities, enabling remote network management via a data communications network (DCN). The DCN consists of connectivity between the management platform (MP) and the payload-carrying network. The DCN includes out-of-band circuits between the MP at the NOC and in-band or integrated circuits on the payload-carrying network. The critical technology involved on the integrated network includes the data communications channel (DCC) for synchronous optical networks (SONET)/ synchronous digital hierarchy (SDH) communications and the optical supervisory channel (OSC) for dense wave division multiplexing (DWDM) communications [HL06].

Optical networks can provide more bandwidth than regular electronic networks can. Major advantages of partial optical networks are gained by incorporating some electronic switching functions. The optical network shown in Figure 5.15 consists of optical nodes interconnected with physical optical links. A point-to-point physical optical link may consist of several logical paths called lightpaths, each carrying a different wavelength. A lightpath is set up by assigning a dedicated wavelength to it. The data can be sent over the lightpath once it is set up. A lightpath can create virtual neighbors out of nodes that may be geographically far apart in the network. A lightpath uses the same wavelength on all fiber links through which it passes.

Optical networks are characterized by a number of advantages over other types of networks. Most of these advantages stem from the use of optical fibers as the communication medium and are summarized below [POP01]:

- Higher bandwidths. Optical fibers offer significantly higher bandwidths as an alternative communication media, offering more than satisfactory performance for the ever-increasing demand on bandwidth. To gain an understanding of the potential of fibers for offering bandwidth, recall that in just the 1.5 μm (1 μm $= 10^{-6}$m) band of each single-mode fiber, the available bandwidth is about 25,000 GHz, three orders of magnitude more than the entire usable radio-frequency bandwidth on Earth.

- Better signal qualities. Since optical transmission is not affected by electromagnetic fields, optical fibers exhibit superior performance than other alternatives, like copper wires. At a given distance the bit error rate (BER) of a fiber-based transmission is significantly better than the BER of a copper or wireless-based transmission. Also, optical fiber provides better noise immunity than the other transmission media, which suffer from considerable electromagnetic interference.

- Easy to deploy and maintain. A good quality optical fiber is sometimes less fragile than a copper-based link. Moreover, optical fibers are not subject to corrosion making them more tolerant to environmental hazards. Furthermore, optical fibers weightless than copper wires making them the technology of choice for deployment over long distances.

- Better security. Optical fiber provides a secure transmission medium since it is not possible to read or change optical signals without physical disruption. It is possible to break a fiber cable and insert a tap, but that involves a temporary disruption. For many critical applications, including military and e-commerce applications, where security is of the utmost importance, optical fibers are preferred over copper transmission media, which can be tapped from their electromagnetic radiations.

As a typical example, the Lucent GNOCs (Global Network Operations Centers) have harnessed all the latest optical networking technological developments to their advantage to deliver state-of-the-art service delivery for next-generation voice, data, and video networks. The model for service delivery is based upon offering standard NOC functions of fault, configuration, asset, performance, and security management. Figure 5.16 [HL06] provides a view of a typical optical network managed by Lucent's GNOC.

An optical network comprise various components. These components can be divided into switching and nonswitching components. Switching components are programmable and enable networking while nonswitching components are used on optical links. Below we briefly summarize these components:

- Optical fibers. Optical fibers constitute the medium for optical signal transmission. There are two basic types of optical fibers, multimode and single mode. Multimode fibers have a wide core, which leads to several

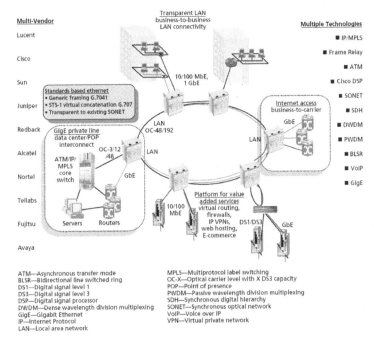

Figure 5.16: Typical Optical Network

rays entering the fiber, with each one traveling over a different path. This fact induces a certain amount of dispersion. Single-mode fibers on the other hand have a significantly narrower core, which leads optical signals to travel over very few paths. This leads to dispersion reduction, thus enabling the transmission of light pulses over greater distances.

- Multiplexers and demultiplexers. These are components that enable the transmission of data streams over a single-fiber channel by using different wavelength perstream.

- Optical transmitters. These are devices that convert electrical signals to optical ones and transmit the latter over the fiber link.

- Optical receivers. These are devices that convert incoming optical signals to electrical ones.

- Optical amplifiers. Amplifiers are devices that amplify the optical signal in order to increase the distance between communicating parties. The performance of optical amplifiers has improved significantly with current amplifiers providing lower noise.

- WDM passive star couplers. These are broadcast devices that that take all their input signals and broadcast them to all outgoing optical paths.

- Optical packet switches. These are devices that provide functionality similar to that of electronic switches by routing an incoming wavelength to a variety of physical output optical paths.

- Wavelength converters. These are devices that convert the wavelength of an incoming signal to a different outgoing wavelength. The conversion takes place entirely in the optical domain.

5.3.2   Management of Optical Networks

Management Challenges of Optical Networks

Network management for optical networks faces some challenges and still unsolved problems.

One of the main premises of optical networks is the establishment of a robust and flexible optical control plane for managing network resources, provisioning, and maintaining network connections across multiple control domains. Such a control plane must have the ability to select lightpaths for requested end-to-end connections, assign wavelengths to these lightpaths, configure the appropriate optical resources in the network, and provide efficient fault management.

The optical network is evolving and being implemented on top of an existing SONET architecture, which provides its own restoration and protection schemes. Without a highly intelligent network management system (NMS), it becomes extremely difficult to ensure that restoration schemes between the electrical and optical layer do not conflict. In addition to mediation between the optical and SONET layer, the network management system must be able to prevent possible conflicts or, at the minimum, enable the service provider to identify conflicts.

In addition to managing the overall network architecture, NMSs must be able to monitor signal performance for each wavelength. With the addition of optical add/drop multiplexers and optical cross-connects, the end-to-end performance of wavelengths becomes more difficult. NMSs for the optical network must assist providers in troubleshooting the network by isolating questionable wavelengths and the possible location of degradation. As the number of wavelengths on each fiber approaches or more, it is important to have an intelligent method to monitor all of them.

Finally, and perhaps most important to the service providers, the ability to manage and provide new services to customers quickly is crucial. Provisioning end-to-end services can be difficult, especially as network capacity decreases. An intelligent NMS can help providers establish and monitor new end-to-end wavelength services to maximize their bandwidth revenues.

Performance management

Performance management is a major complication in optical networks, since optical performance measurements, which are typically limited to optical power, optical signal-to-noise ratio (OSNR), and wavelength registration, do not directly relate to QoS measures used by carriers. These are concerned with attributes related to the lightpath, such as BER and parity checks, of which the management system may have no prior knowledge. Moreover, transparency means that it is not possible to access overhead bits in the transmitted data to obtain performance related measures, adding further complexity to the detection of service disruption. Unless information concerning the type of signal that is being carried on a lightpath is conveyed to the NMS, it will not be able to ascertain whether the measured power levels and OSNR fall within the preset acceptable limits.

Routing control

In wavelength-routed optical networks there must exist a control mechanism that governs the establishment and release of optical connections. Upon the reception of a connection request such a mechanism will select an appropriate route, assign the appropriate wavelength to the connection, and configure the networks optical switches accordingly. This mechanism is also responsible for keeping track of which wavelengths are currently used in each optical link in order to enable correct routing decisions. Such a control mechanism can be either centralized or distributed, with distributed mechanisms being preferable due to their increased robustness. The research in this area has four main goals [POP01]:

- decrease the blocking probability of connection requests,

- decrease connection set-up delays,

- decrease the bandwidth used for control messages,

- increase system scalability.

There exist two distributed network control mechanisms. These are the "link-state approach" and the "distributed routing approach":

- The link-state approach demands that each node maintains information regarding the topology of the complete network and the use of wavelengths at each fiber link. When a connection request arrives, the node uses this information in order to select a route and a wavelength. When those are selected, the node tries to reserve the wavelength along each fiber link on the route by making reservation requests to the nodes on that route. If all intermediate nodes accept the reservation requests, the connection is established, whereas in case of connection failure, all intermediate nodes are notified in order to release the reserved resources. Upon the establishment of a connection, network nodes are notified for the new status of wavelength availability at the nodes of the established route.

- The distributed routing approach selects routes without knowledge of the complete network topology. This approach demands that each node stores a routing table (using a distributed Bellman-Ford algorithm) that contains information regarding the next hop and the associated cost of the shortest path to each destination over a different wavelength. Upon reception of a connection request, a node will: (a) select the wavelength that minimizes the distance to the destination and (b) forward the request to the next node on this path. This procedure forwards the connection request hop-by-hop with the appropriate wavelength being chosen at each hop. Obviously, the connection is established when the request reaches the destination node, which returns an acknowledgement to the source node in order to initiate data transmission. If, however, the connection cannot be established, the nodes along the path are notified to release the reserved wavelengths. Upon establishment of a connection, each node notifies it neighbors about the status of the reserved link and wavelength in order to let the neighbors update their routing tables.

Furthermore, an important issue is how to address the trade-off between service quality and resource utilization. Addressing this issue requires different scheduling and sharing mechanisms to maximize resource utilization while ensuring adequate QoS guarantees. One possible solution is the aggregation of traffic flows to maximize the optical throughput and reduce operational and capital costs.

Another related issue arises from the fact that the implementation of a control plane requires information exchange between control and management entities involved in the control process. To achieve this, fast signaling channels need to be in place between switching and routing nodes. These channels might be used to exchange up-to-date control information needed for managing all supported connections and performing other control functions. In general, control channels can be realized in different ways; one might be implemented in-band, while another may be implemented out-of-band. There are, however, compelling reasons for decoupling control channels from their associated data links.

Fault management

Fault management is further complicated since detection functions, which should be handled at the ISO layer closest to the failure, are delegated to the physical layer instead of higher layers. That is, fault detection and localization methods are less insulated from details of physical layer than of higher layers, requiring the availability of expert diagnostic techniques to measure and control the smallest granular component, the wavelength channel.

There are several approaches to combate network component failures, and are divided in two categories:

- Dedication of backup resources in advance. In this case, network connections are restored by utilizing backup resources that have been reserved for these connections in advance. The advantages of this approach are obviously the fast restoration time and the guaranteed restoration capability. The "1+1 protection" method belongs to this category. According to this method, for each connection that requires protection a dedicated route and wavelength is reserved in advance. When the link fails on the primary connection, the communicating nodes utilize the back-up resources reserved for the link.

- Dynamic restoration. In this case, network operation is restored by utilizing the remaining capacity of the network. The advantages of this approach are the efficient use of network resources and resilience against many kinds of failures. The "link restoration" and "path restoration" methods belong to this category. In the "link restoration" scheme, all the connections that go through a failed element are rerouted around this element. The end nodes of the failed link are responsible for finding a way around the failed link. Upon the establishment of the new link, the end nodes of the failed link configure their routing tables, accordingly. On the other hand, in "path restoration" the failed link is replaced by a new one that is discovered on an end-to-end basis by the pair of communicating nodes and the network elements configure accordingly.

For security management, security failure identification of both the location and type of attack differs significantly from that in traditional networks, which basically relies on identifying the domain of an alarm and using algorithms to determine the probability of a certain failure having occurred. Although the same techniques can be applied to optical networks, several issues exist and need to be addressed carefully.

5.4 Overlay Networks

An overlay network is a computer network that is built on top of another network. Nodes in the overlay can be thought of as being connected by virtual or logical links, each of which corresponds to a path, perhaps through many physical links, in the underlying network. For example, many peer-to-peer networks are overlay networks because they run on top of the Internet. Dial-up Internet is an overlay upon the telephone network.

Overlay networks can be constructed in order to permit routing messages to destinations not specified by an IP address. For example, distributed hash tables can be used to route messages to a node having specific logical address, whose IP address is not known in advance.

Overlay networks have also been proposed as a way to improve Internet routing, such as through quality of service guarantees to achieve higher-quality streaming media. Previous proposals such as IntServ, DiffServ, and IP Multicast have not seen wide acceptance largely because they require modification of all routers in the network. On the other hand, an overlay network can be incrementally deployed on end-hosts running the overlay protocol software, without cooperation from ISPs. The overlay has no control over how packets are routed in the underlying network between two overlay nodes, but it can

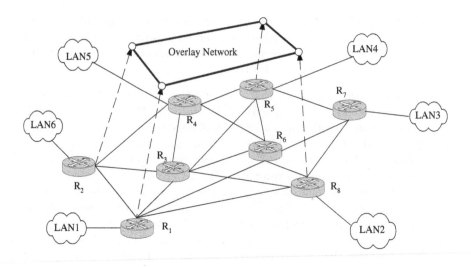

Figure 5.17: An Overlay Network for Connections Between LANs

control, for example, the sequence of overlay nodes a message traverses before reaching its destination.

Overlay networks create a virtual topology on top of an existing physical topology on a public network. Overlay networks are self-organized; thus, if a node fails, the overlay network algorithm can provide solutions that let the network recreate an appropriate network structure.

Figure 5.17 shows an overlay network configured over a wide area network. Nodes in an overlay network can be thought of as being connected by logical links. In Figure 5.17, routers R_1, R_2, R_5, and R_8 are participating in creating an overlay network where the interconnection links are realized as overlay logical links. Such a logical link corresponds to a path in the underlying network. An obvious example of these networks is the peer-to-peer network, which runs on top of the Internet. VPN (Virtual Private Network) is another example, which is on top of public networks. Overlay networks have no control over how packets are routed in the underlying network between a pair of overlay source/destination nodes. However, these networks can control a sequence of overlay nodes through a message-passing function before reaching the destination.

Overlay networks are self-organized. When a node fails, the overlay network algorithm should provide solutions that let the network recover and recreate an appropriate network structure. Another fundamental difference between an overlay network and an unstructured network is that overlays' look-up routing information is on the basis of identifiers derived from the content of moving frames.

5.4.1   Management of Peer-to-Peer Networks

Introduction of Peer-to-Peer Networks

Peer-to-peer (P2P) networks are distributed systems where the software running at each node provides equivalent functions. A succinct formal definition of P2P networking is "a set of technologies that enable the direct exchange of services or data between computers." Implicit in that definition are the fundamental principles that peers are equals. P2P systems emphasize sharing among these equals. A pure peer-to-peer system runs without any centralized control or hierarchical organization. A hybrid system uses some centralized or hierarchical resources. Peers can represent clients, servers, routers, or even networks [HB05].

Figure 5.18 illustrates simple reference architectures for both client-server systems and P2P systems. Clients are not equal to servers, and they depend on a relatively small number of servers for system operation. Peers are all equal, and they rely only on themselves and their peers.

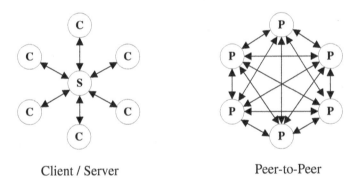

Client / Server   Peer-to-Peer

Figure 5.18:   Client/Server Model and P2P Model

An important goal in P2P networks is that all clients provide resources, including bandwidth, storage space, and computing power. Thus, as nodes arrive and demand on the system increases, the total capacity of the system also increases. This is not true of a client-server architecture with a fixed set of servers, in which adding more clients could mean slower data transfer for all users.

The distributed nature of P2P networks also increases robustness in case of failures by replicating data over multiple peers, and, in pure P2P systems, by enabling peers to find the data without relying on a centralized index server. In the latter case, there is no single point of failure in the system.

P2P networks can be classified by what they can be used for:

- file sharing

- telephony

- media streaming (audio, video)

- discussion forums

Other classification of P2P networks is according to their degree of centralization. In "pure" P2P networks, 1)Peers act as equals, merging the roles of clients and server; 2) There is no central server managing the network; 3) There is no central router.

Architectures of Peer-to-Peer Networks

Every peer-to-peer network uses one of three types of architectural formats. These formats may include peers and servers. A peer can be a user's computer, running a P2P client. These workstations enact searches for files, provide files for upload and download, and send files to fulfill requests. A server provides features for peers to enter available files, compiles lists of the available files, responds to searches received from peers, and, depending on the architecture, aids in uploading and downloading files through firewalls.

- Centralized Architecture: Central server responds to peer requests

- De-centralized Architecture: Multiple peers respond to requests from other peers on the network

- Hybrid Architecture: Multiple servers (SuperNodes) respond to requests by communicating with servers on the network, compiling and sending responses through the primary server first queried

1. Centralized Architecture

In this architecture, the peer-to-peer application executing on the peer systems establishes a persistent connection to the central server. Users log into this central server to access the network. The peer system transmits a directory listing of all available items for sharing and downloading. See Figure 5.19.

The Central Server maintains a database of all shared items. When the server receives requests, it responds with a listing of available matches and contact information of the host, such as an IP address and port number. When a user selects an item to download, the downloading peer contacts the hosting peer directly, transferring the file peer to peer. In sophisticated networks, the peer-to-peer application sends a unique identifier (such as a hash number) for each shared item. The central server sends the peer-to-peer application a list of peers hosting identical items. In this architecture, the peer-to-peer application establishes connections to multiple peers and downloads sections of the file simultaneously, which the P2P application reassembles.

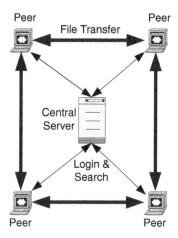

Figure 5.19: Centralized Architecture of P2P

The centralized architecture provides excellent performance for search requests and is popular in smaller networks where the community controls user access. However, the centralized architecture does not scale adequately to large networks and suffers a severe weakness with the central server. Hackers and malicious attacks can easily disable peer-to-peer Networks built on the centralized architecture by attacking and disabling the central server.

2. Decentralized Architecture

The decentralized architecture uses a distributed computing model in which each peer is an equal within the network. The decentralized architecture does not contain a central server, which purist administrators would consider a "true" peer-to-peer network. In this architecture, the P2P application executing on peer systems establishes persistent connections to peers within the network. See Figure 5.20.

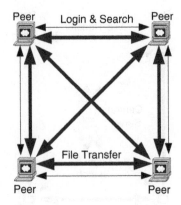

Figure 5.20: Decentralized Architecture of P2P

The peer system sends search requests to each of the persistent connections on the network, broadcasting out from the central connection. Matches to the search request return to the requestor from each peer, detailing contact information (such as IP address and port number) for hosting peer.

When a user selects an item to download, the downloading peer contacts the hosting peer directly. In response, the host transfers the file between the two peers. As in the Centralized Architecture, advanced peer-to-peer applications can establish connections to multiple peers and download sections of the file simultaneously from the multiple hosts. When complete, the P2P application reassembles the file.

The decentralized architecture offers two primary advantages over the centralized approach. First, this architecture scales to large networks of peers. Secondly, malicious attackers cannot easily disable the decentralized approach because of the distributed control. The disadvantage to decentralized networks is the significantly longer time required to perform search operations.

3. Hybrid Architecture

The hybrid architecture combines the centralized and decentralized approaches into one architecture. The hybrid architecture introduces the concept of a SuperNode (also known as an UltraPeer). The SuperNode functions in a similar function to the central server of the Centralized Architecture. In this architecture, SuperNodes are geographically dispersed to create a larger network. The peer-to-peer application executing on the peer systems establishes a persistent connection to one or more SuperNode(s) and transmits a directory listing of the items available for sharing on the peer system. See Figure 5.21.

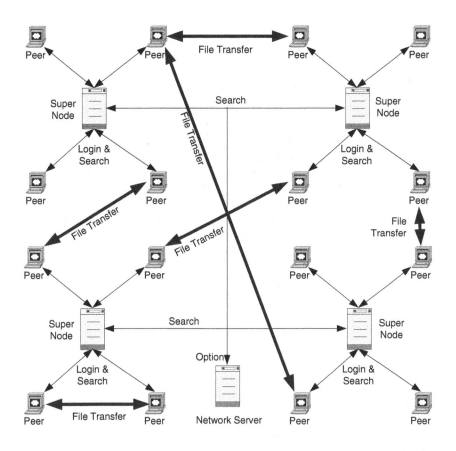

Figure 5.21: Hybrid Architecture of P2P

Each SuperNode maintains a database of shared items. The P2P application sends requests to the SuperNodes, which forward to additional Super-Nodes. The primary server compiles the responses and sends the peer a list of matches and host contact information (such as IP address and port number). When a user selects an item to download, the downloading peer contacts the hosting peer directly and transfers the file between the two peers. As in the other architectures, advanced peer-to-peer applications can establish connections to multiple peers and download sections of the file simultaneously from the multiple hosts. The peer-to-peer application reassembles the sections into a complete file.

Typically, the SuperNode does not transfer the file between the peers. However, as with the Centralized Architecture, the SuperNode aids connections and transfers when the hosting peer resides behind corporate or personal firewalls. In this architecture, the SuperNode sends a command to the hosting peer to connect to the downloading peer. The downloading peer contacts the hosting peer directly with the aid of the SuperNode and transfers the file between the two peers.

Compared to central servers, SuperNodes are more dynamic. These servers are typically "promoted" from a peer that has a fast CPU, high bandwidth access to the Internet, and capable of supporting 200 - 300 simultaneous connections. SuperNodes maintain a list of available SuperNodes and transmit this list on a regularly to all peers connected to the P2P network. Peer systems caches the information, loading the updated link lists during startup.

Another option for Hybrid networks employs a network portal server into the architecture. The network portal, typically owned and provided by the "owner" of the network, hosts web services for the P2P network. Services can include a home page with news, forums, chat, or instant messaging options and typically functions as an advertisement server to deliver ads to the peer-to-peer clients, thereby generating revenue for the owner of the network. The network server may also act as a registration server for users to log into the network and distributes the initial SuperNode IP addresses to new peer-to-peer clients. The hybrid architecture offers the best of both the centralized and the decentralized approaches. Like the Centralized Architecture, the Hybrid provides excellent performance for search requests, even in a large distributed network.

The hybrid architecture scales to large networks of peers. As with the decentralized approach, the hybrid network cannot be easily disabled due to the distributed and dynamic nature of the SuperNodes.

Management of Peer-to-Peer Networks

1. Traffic management

The primary obstacle in managing peer-to-peer applications is the detection of peer-to-peer traffic. The requirements for traffic management in P2P networks are as follows:

- Ensure that critical applications are not impacted by nonpriority traffic
- Deliver optimal application performance by allocating more bandwidth for higher priority applications
- Eliminate special purpose Rate Shaping products for simplified, centralized traffic management capabilities
- Provide flexible bandwidth limits, bandwidth borrowing, and traffic queuing
- Control rate classes based on any traffic variable
- Enable application bandwidth to be shared across similar priority applications for better resource sharing

- Ensure that specific types of application traffic stay within authorized boundaries

Most current versions of peer-to-peer applications are utilizing an architecture that involves SuperNodes or Hubs. The peers typically create a persistent TCP connection to the SuperNode. This peer then performs logins, searches, and other vital functions using the persistent connection. The ability to detect and manage the connection(s) from the peer to the SuperNode provides great control over the peer-to-peer network. And in turn, blocking the network traffic on this connection effectively disables the peer-to-peer application.

File transfers are identified as two types: GET and PUT (or SEND). A peer typically requests a file from another peer by issuing some form of GET request. In other situations, a PUT command issues to a peer to instruct the application to perform an out-bound connection to another peer and then transfer the file. In some cases, the GET/PUT command occurs over the persistent Peer-to-SuperNode connection, or the command occurs over the peer-to-peer connection. In the case where the request occurs on the peer-to-peer connection, the file transfer can be blocked or rate-limited. In the case where the request occurs on the Peer-to-SuperNode connection, the file transfer can only be blocked.

Some techniques are used for rate-limiting and get to the goal of traffic control.

2. Self-Organization in Peer-to-Peer

Peer-to-peer systems have to provide services like routing and searching for and accessing of resources. An open question is if and how much can self-organization, with all its illusiveness, emerge as an essential means for improving the quality of the services. Improved service quality, thereby, is to be achieved equally for performance, robustness, security, and scalability in an all open world.

Requirements for self-organization in peer-to-peer networks [MK05]:

- Feedback:

  A self-organizing peer-to-peer system is often exposed to positive and negative feedback, whereupon the structure or behavior of the system changes in a balanced way. Feedback includes messages that peers send to each other.

- Reduction of complexity:

  A peer-to-peer system that is self-organizing develops structures and hides details from the environment to reduce the overall complexity. This may include forming clusters or creating creation of other entities.

- Randomness:

A self-organizing system makes use of randomness as a prerequisite for creativity. This allows the creation of new structures with little effort. An example from another discipline are ant algorithms where ants decide randomly between different paths when they have no sufficient knowledge about their environment.

- Self-organized criticality (SOC):

  A system that is self-organizing drives itself into a state of criticality. Too much order as well as too much disorder are to be avoided by adequate procedures. This should result in an increased degree of flexibility because the system is able to cope with different types of perturbations.

- Emergence:

  A self-organizing peer-to-peer system shows properties that no single peer has on its own, or properties that may have been unknown at design time. Peers forming a small-world exemplify an emergent structure.

Self-organization is seen as an attractive feature of peer-to-peer networks as it essentially enables running a complex system without exercising stricter form of control and management.

3. Security Management

Many peer-to-peer networks are under constant attack such as:

- poisoning attacks (e.g., providing files whose contents are different from the description)

- polluting attacks (e.g., inserting "bad" chunks/packets into an otherwise valid file on the network)

- defection attacks (users or software that make use of the network without contributing resources to it)

- insertion of viruses to carried data (e.g., downloaded or carried files may be infected with viruses or other malware)

- malware in the peer-to-peer network software itself (e.g., distributed software may contain spyware)

- denial of service attacks (attacks that may make the network run very slowly or break completely)

- filtering (network operators may attempt to prevent peer-to-peer network data from being carried)

- identity attacks (e.g., tracking down the users of the network and harassing or legally attacking them)

- spamming (e.g., sending unsolicited information across the network not necessarily as a denial of service attack)

Most attacks can be defeated or controlled by careful design of the peer-to-peer network and through the use of encryption. P2P network defense is in fact closely related to the "Byzantine Generals Problem." However, almost any network will fail when the majority of the peers are trying to damage it, and many protocols may be rendered impotent by far fewer numbers.

5.4.2 Management of VPN (Virtual Private Networks)

Introduction of VPN

A virtual private network (VPN) is a computer network in which some of the links between nodes are carried by open connections, or virtual circuits, in some larger network (e.g., the Internet) as opposed to their conduction across a single private network. The link-layer protocols of the virtual network are said to be tunneled through the larger network when this is the case. One common application is secure communications through the public Internet, but a VPN need not have explicit security features, such as authentication or content encryption. VPNs can be used to separate the traffic of different user communities over an underlying network with strong security features.

A VPN may have best-effort performance or may have a defined service level agreement (SLA) between the VPN customer and the VPN service provider. Generally, a VPN has a topology more complex than point to point. A VPN allows computer users to access a network via an IP address other than the one that actually connects their computer to the Internet.

The Internet Engineering Task Force (IETF) categorized a variety of VPNs, some of which, such as Virtual LANs (VLAN) are the standardization responsibility of other organizations, such as the Institute of Electrical and Electronics Engineers (IEEE) Project 802, Workgroup 802.1 (architecture). Originally, network nodes within a single enterprise were interconnected with Wide Area Network (WAN) links from a telecommunications service provider. With the advent of LANs, enterprises could interconnect their nodes with links that they owned. While the original WANs used dedicated lines and layer 2 multiplexed services such as Frame Relay, IP-based layer 3 networks, such as the ARPANET, Internet, military IP networks, became common interconnection media. VPNs began to be defined over IP networks.

When an enterprise interconnected a set of nodes, all under its administrative control, through an LAN network, that was termed an Intranet. When the interconnected nodes were under multiple administrative authorities, but were hidden from the public Internet, the resulting set of nodes was called an Extranet. Both Intranets and Extranets could be managed by a user organization, or the service could be obtained as a contracted offering, usually customized, from an IP service provider. In the latter case, the user organization contracted for layer 3 services much as it had contracted for layer 1 services such as dedicated lines, or multiplexed layer 2 services such as frame relay.

The IETF distinguishes between provider-provisioned and customer-provisioned VPNs. Much as conventional WAN services can be provided by an interconnected set of providers, provider-provisioned VPNs (PPVPNs) can be provided by a single service provider that presents a common point of contact to the user organization.

MPLS networks are a good example of VPNs. In MPLS, multiple labels can be combined in a packet to form a header for efficient tunneling. The Label Distribution Protocol (LDP) is a set of rules by which routers exchange information effectively. MPLS uses traffic engineering for efficient link bandwidth assignments. Such networks operate by establishing a secure tunnel over a public network. Finally, we look at overlay networks. An overlay network is a computer network that creates a virtual topology on top of the physical topology of the public network.

VPNs are deployed with privacy through the use of a tunneling protocol and security procedures. Figure 5.22 shows two organizations, 1 and 4, connected through their corresponding routers, forming a tunnel in the public network, such as the Internet. Tunneling has two forms: remote-access tunneling, which is a user-to-LAN connection, and site-to-site tunneling, whereby an organization can connect multiple fixed sites over a public network. Such

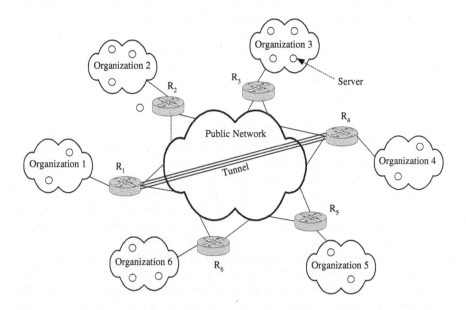

Figure 5.22: An Example of VPN Through a Tunnel Using Public Facilities

a structure gives both private organizations the same capabilities they have on their own networks but at much lower cost. They can do this by using the shared public infrastructure. Creating a VPN benefits an organization benefits by providing:

- Extended geographical communication

- Reduced operational cost

- Enhanced organizational management

- Enhanced network management with simplified local area networks

- Improved productivity and globalization

Management of VPN

With the globalization of businesses, many companies have facilities across the world and use VPNs to maintain fast, secure, and reliable communications across their branches. But the management of VPN meets some challenges.

Security Management

VPNs remain susceptible to security issues when they try to connect between two private networks using a public resource. The challenge in making a practical VPN, therefore, is finding the best security for it.

From the security standpoint, either the underlying delivery network is trusted, or the VPN must enforce security with mechanisms in the VPN itself. Unless the trusted delivery network runs only among physically secure sites, both trusted and secure models need an authentication mechanism for users to gain access to the VPN.

Some ISPs now offer managed VPN service for business customers who want the security and convenience of a VPN but prefer not to undertake administering a VPN server themselves. Managed VPNs go beyond VPN scope and are a contracted security solution that can reach into hosts. In addition to providing remote workers with secure access to their employer's internal network, other security and management services are sometimes included as part of the package. Examples include keeping anti-virus and anti-spyware programs updated on each client's computer.

Authentication before VPN connection

There are a wide variety of authentication mechanisms, which may be implemented in devices including firewalls, access gateways, and other devices. They may use passwords, biometrics, or cryptographic methods. Strong authentication involves combining cryptography with another authentication mechanism. The authentication mechanism may require explicit user action, or may be embedded in the VPN client or the workstation.

Trusted Delivery Networks

Trusted VPNs (sometimes referred to APNs, or Actual Private Networks) do not use cryptographic tunneling, and instead rely on the security of a single provider's network to protect the traffic. In a sense, these are an elaboration of traditional network and system administration work.

Multi-Protocol Label Switching (MPLS) is often used to overlay VPNs, often with quality of service control over a trusted delivery network.

Layer 2 Tunneling Protocol (L2TP), which is a standards-based replacement, and a compromise taking the good features from each, for two proprietary VPN protocols: Cisco's Layer 2 Forwarding (L2F) (now obsolete) and Microsoft's Point-to-Point Tunneling Protocol (PPTP).

Security Mechanisms

Secure VPNs use cryptographic tunneling protocols to provide the intended confidentiality (blocking snooping and thus Packet sniffing), sender authentication (blocking identity spoofing), and message integrity (blocking message alteration) to achieve privacy. When properly chosen, implemented, and used, such techniques can provide secure communications over unsecured networks.

Secure VPN protocols include the following:

- IPsec (IP security) commonly used over IPv4, and a "standard option" in IPv6.

- SSL/TLS used either for tunneling the entire network stack, as in the OpenVPN project, or for securing what is, essentially, a web proxy. SSL is a framework more often associated with e-commerce, but it has been built upon by a number of vendors to provide remote access VPN capabilities. A major practical advantage of an SSL-based VPN is that it can be accessed from the locations that restrict external access to SSL-based e-commerce websites only, thereby preventing VPN connectivity using IPsec protocols. SSL-based VPNs are vulnerable to trivial Denial of Service attacks mounted against their TCP connections because latter are inherently unauthenticated.

- OpenVPN, an open standard VPN. It is a variation of SSL-based VPN that is capable of running over UDP. Clients and servers are available for all major operating systems.

- L2TPv3 (Layer 2 Tunneling Protocol version 3), a new release.

- VPN Quarantine. The client machine at the end of a VPN could be a threat and a source of attack; this has no connection with VPN design and is usually left to system administration efforts. There are solutions that provide VPN Quarantine services, which run end point checks on the remote client while the client is kept in a quarantine zone until healthy.

Security on Mobile VPNs

Mobile VPNs are VPNs designed for mobile and wireless users. They integrate standards-based authentication and encryption technologies to secure data transmissions to and from devices and to protect networks from unauthorized users.

Designed for wireless environments, Mobile VPNs are designed as an access solution for users that are on the move and require secure access to information and applications over a variety of wired and wireless networks. Mobile VPNs allow users to roam seamlessly across IP-based networks and in and out of wireless coverage areas without losing application sessions or dropping the secure VPN session. For instance, highway patrol officers require access to mission-critical applications in order to perform their jobs as they travel across different subnets of a mobile network, much as a cellular radio has to hand off its link to repeaters at different cell towers.

Host Identity Protocol (HIP) is an IETF protocol supporting mobile VPN using Bound End-to-End Tunnel (BEET) mode for IPsec.

Service Management

VPN Management simplifies the task of globally defining, distributing, enforcing, and deploying VPN policies for managed VPN gateways to keep all remote sites in synch with the latest security policies.

VPN Management delivers a cost-effective global management solution that reduces staffing requirements, speeds up deployment, and lowers the cost of delivering services by centralizing the management security policies.

Data Management

VPN Management supports access to industry-leading, back-end databases for highly efficient and reliable data storage and retrieval. It stores its configuration files and data in databases to automatically be backed up and restored.

Tunnel Management can be thought of in terms of what tunneling protocol(s) will be used, network subnetting, and bandwidth absorption. It is imperative to plan what protocol(s) will be used in the VPN. There are "legacy tunnel protocols" (such as PPTP) as well as "new technology tunnel protocols" (such as IPSec) to chose from. Also complicating tunnel management is the area of network subnetting. Planning, researching, reporting, and ultimately deploying a VPN are important management tasks in VPN applications. The effect of tunnels is correlated with network bandwidth.

5.5　Grid Architectures

5.5.1　Introduction of Grid Networks

Grid computing is a computing model that provides the ability to perform higher throughout computing by taking advantage of many networked computers to model a virtual computer architecture. Grids use the resources of many separate computers connected by a network to solve large-scale computation problems. Grids provide the ability to perform computations on large data sets, by breaking them down into many small ones, or provide the ability to perform many more computations at once than would be possible on a single computer, by modeling a parallel division of labor between processes.

The Grid provides scalable, reliable, secure mechanisms for discovering, assembling, integrating, utilizing, reconfiguring, and releasing multiple heterogeneous resources. These resources can include compute clusters, specialized computers, software, mass storage, data repositories, instruments, and sensors. However, Grid environments are not restricted to incorporating common information technology resources. The architecture is highly expandable and can extend to virtually any type of device with a communications capability.

Grid architecture has been designed specifically to enable it to be used to create many different types of services. A design objective is the creation of infrastructure that can provide sufficient levels of abstraction to support an almost unlimited number of specialized services without the restrictions of dependencies inherent in delivery mechanisms, local sites, or particular devices. These environments are designed to support services not as discrete infrastructure components, but as modular resources that can be integrated into specialized blends of capabilities to create multiple additional, highly customizable services. The Grid also allows such services to be designed and implemented by diverse, distributed communities, independently of centralized processes. Grid architecture represents an innovation that is advancing efforts to achieve these goals.

The Grid is a major new type of infrastructure that builds upon, and extends the power of, innovations that originally arose from addressing the requirements of large-scale, resource-intensive science and engineering applications.

These requirements, as well as multi-year development implementation cycles, lead to architectural designs that eliminate dependencies on particular hardware and software designs and configurations. Scientists designing and developing major applications cannot become dependent either on static infrastructure, given the rate of technology change, or on infrastructure that is subject to continual changes in basic architecture. They require a degree of separation between their applications and specific, highly defined hardware and configurations.

Although the Grid was initially developed to support large-scale science projects, its usefulness became quickly apparent to many other application communities. The potential of its architecture for abstracting capabilities from underlying infrastructure provide a means to resolve many issues related to information technology services.

The general features of Grids

General Grid characteristics include the following features. Each of these features can be formally expressed within an architectural framework. Within Grid environments, to a significant degree, these determinations can be considered more art than craft. Ultimately, it is the application or service designer who can determine the relationship among these functions [TMK06].

- Abstraction/virtualization. Grids have exceptional potential for abstracting limitless customizable functions from underlying information technology infrastructure and related resources. The level of abstraction within a Grid environment enables support for many categories of innovative applications that cannot be created with traditional infrastructure, because it provides unique methods for reducing specific local dependencies and for resource sharing and integration.

- Resource sharing. One consequence of this support for high levels of abstraction is that Grid environments are highly complementary to services based on resource sharing.

- Flexibility/programmability. Another particularly important characteristic of the Grid is that it is a "programmable" environment, in the sense of macroprogramming and resource steering. This programmability is a major advantage of Grid architecture: providing flexibility not inherent in other infrastructure, especially capabilities made possible by workflow management and resource reconfigurability. Grids can enable scheduled processes and/or continual, dynamic changing of resource allocations and configurations, in real time. Grids can be used to support environments that require sophisticated orchestration of workflow processes. Much of this flexibility is made possible by specialized software "toolkits," middleware that manages requests and resources within workflow frameworks.

- Determinism. Grid processes enable applications to directly ensure, through autonomous processes, that they are matched with appropriate service levels and required resources, for example, through explicit signaling for specialized services and data treatments.

- Decentralized management and control. Another key feature underlying Grid flexibility is that its architecture supports the decentralization of management and control over resources, enabling multiple capabilities to be evoked independently of processes that require intercession by centralized processes.

- Dynamic integration. Grids also allow for the dynamic creation of integrated collections of resources that can be used to support special higher-level environments, including such constructs as virtual organizations.

- Resource sharing. Grid abstraction capabilities allow for large-scale resource sharing among multiple, highly distributed sites.

- Scalability. Grid environments are particularly scalable. They can be implemented locally or distributed across large geographic regions, enabling the reach of specialized capabilities to extend to remote sites across the world.

- High performance. Grids can provide for extremely high-performance services by aggregating multiple resources, e.g., multiple distributed parallel processors and parallel communication channels.

- Security. Grids can be highly secure, especially when segmentation techniques are used to isolate partitioned areas of the environment.

- Pervasiveness. Grids can be extremely pervasive and can extend to many types of edge environments and devices.

- Customization. Grids can be customized to address highly specialized requirements, conditions, and resources.

Extending general Grid attributes to communication services and network resources has been an evolutionary process. A key goal has been to ensure that these services and resources can be closely integrated with multiple other co-existent Grid services and resources. This close integration is one of the capabilities that enable networks to become "full participants" within Grid environments, as opposed to being used as generic, accessible external resources.

The Internet design has been a major benefit to Grid deployments. Unlike legacy telecommunications infrastructure, which has had a complex core and minimal functionality at the edge, the Internet places a premium on functionality at the edge supported by a fairly simple core. This end-to-end design principle enables innovation services to be created and implemented at the edge of the network, provides for high-performance network backbones, and allows for significant service scalability.

5.5.2 Management of Grid Networks

Resource Management

A primary motivation for the design and development of Grid architecture has been to enhance capabilities for resource sharing, for example, utilizing spare computation cycles for multiple projects. Similarly, a major advantage to Grid networks is that they provide options for resource sharing that are difficult if not impossible in traditional data networks. Visualization of network resources allows for the creation of new types of data networks, based on resource sharing techniques that have not been possible to implement until recently.

The Grid development communities are engaged in implementing Grid infrastructure software with Web Services components. These components provide access to sets of building blocks that can be combined easily into different service combinations within classes, based on multiple parameters. They can be used to customize services and also to enable shared resources within autonomous environments.

Within a Grid network services context, these capabilities provide new mechanisms for network services design and provisioning, especially new methods for directly manipulating network resources. This approach allows for the creation of customized services by integrating different services at different network layers, including through interlayer signaling, to provide precise capabilities required by categories of applications that cannot be deployed, or optimized, within other environments. Using these techniques, novel network services can be based on multiple characteristics, e.g., those based on policy-based access control and other forms of security, priority of traffic flows, quality of service guarantees, resource allocation schemes, traffic shaping, monitoring, prefault detection adjustments, and restoration techniques.

The resource management system must cope with additional tasks in cluster management systems:

- the support of jobs coming from various sources

- the interaction with different cluster management systems

- the provision of additional services for the grid layer

EU DataGrid project [RSR+04] developed a single component, an abstraction layer, that facilitates the deployment of advanced schedulers and provides a consistent interface to different cluster management systems.

Decentralized Management

Decentralized control and management of resources allows resource provisioning, utilization, and reconfiguration without intercession by centralized management or other authorities. During the past few years, various technologies and techniques have been developed to allow decentralized control over network resources. These methods allow Grid networks to be "programmed," significantly expanding Grid network services capabilities. Today, methods are available that can provide multiple levels of deterministic, differentiated services capabilities not only for layer 3 routing but also for services at all other communication layers.

Some of these methods are based on specialized signaling, which can be implemented in accordance with several basic models. For example, two basic models can be considered two ends of a spectrum. At one end is a model based on predetermining network services, conditions, and attributes, and providing service qualities in advance, integrated within the core infrastructure. At the other end is a model based on mechanisms that continually monitor network conditions and adjust network services and resources based on those changes. Between these end points, there are techniques that combined preprovisioning methods with those based on dynamic monitoring and adjustment. Emerging Grid networking techniques define methods that provide for determinism by allowing applications to have precision control over network resource elements when required.

Grid environments are by definition highly distributed and are, therefore, highly scalable geographically. Consequently, Grid networks can extend not only across metro areas, regions, and nations but also worldwide. The scalability of advanced Grid networks across the globe has been demonstrated for the past several years by many international communities, particularly those using international networks.

Currently, the majority of advanced Grid networks are being used to support global science applications on high-performance international research and education networks. This global extension of services related to these projects has been demonstrated not only at the level of infrastructure but also with regard to specialized services and dynamic allocation and reconfiguration capabilities. EU project GridCC (Grid Enabled Remote Instrumentation with Distributed Control and Computation) is a typical application example in this erea [DMP+08].

Autonomic Management

The autonomic management of all hardware and software components in Grids includes [RSR+04]:

- Automatically adding new machines to a fabric, i.e., recognizing those machines, installing appropriate software packages on them and configuring the services hosted by a machine.

  Assuming a new machine has been added or an existing one has been upgraded, the following actions are performed: (1) installing/running the core operating system and the self-management components, (2) deciding which services the machine should host, (3) updating the machine's configuration accordingly (installing software packages and configuring the services running on the machine). Both steps (1) and (2) can be initiated from the machine itself via some bootstrapping mechanism or externally from another machine that observes the network to recognize new machines.

- Ensuring the functional integrity of hardware and software components, including to switch off faulty services or even removing hosts from the active set of machines.

A Grid computing center is build of many services such as a file server, a DNS server, a Grid access node, a batch master, a batch worker, etc. The relationship of services to machines (hardware) is typically:

- some services exist only once (e.g., DNS server),
- multiple machines host a single specific service only (e.g., file server and batch worker),
- some machines host multiple services (e.g., Grid access and batch master).

The self-management must ensure that the services are running and function correctly.

Performance Management

Because many Grid applications are extremely resource intensive, one of the primary drivers for Grid design and development has been the need to support applications requiring ultra-high-performance data computation, flow, and storage. Similarly, Grid networks require extremely high-performance capabilities, especially to support data-intensive flows that cannot be sustained by traditional data networks. Many of the current Grid networking research and development initiatives are directed at enhancing high-performance data flows, such as those required by high-energy physics, computational astrophysics, visualization, and bioinformatics. For Grid networks, high performance is measured by more than support for high volume data flows. Performance is also measured by capabilities for fine-grained application control over individual data flows. In addition, within Grid networks, performance is also defined by many other measures, including end-to-end application behavior, differentiated services capabilities, programmability, precision control responsiveness, reconfigurability, fault tolerance, stability, reliability, and speed of restoration under fault conditions.

Job Management

User jobs in Grids may originate from different sources: local user's community, Grid community, or self-management part.

Because certain self-management actions could corrupt the run-time environment of jobs, the job management must provide capabilities to coordinate user jobs with maintenance jobs. Different Grid computing centers may deploy different job management systems each providing a specific set of scheduling features apart from common functionality. Hence, job management must provide methods to enhance scheduling features.

Jobs from the Grid community must pass an authorization check before they can be submitted to a local queuing system. Authorization may depend on multiple factors, e.g., user's affiliation, resource requirements, current state of the resources, and may change dynamically. If a job request has passed the authorization step it must be provided with the necessary credentials for their execution.

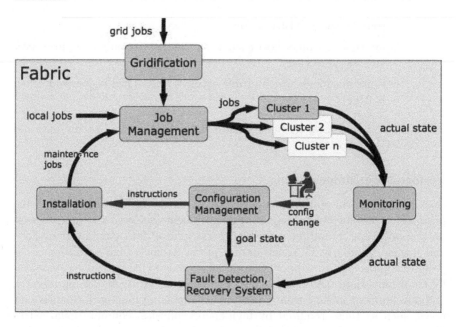

Figure 5.23: Job Management in a Grid

Figure 5.23 [RSR+04] depicts the interrelationships between the components in example job management and shows how they work together to achieve an automated computing center management. A "fabric" comprises one to several separate clusters with (possibly) different cluster management systems (batch systems) and a couple of servers responsible for services, e.g., self-management components, file servers, DNS server, Grid access node, etc. (not shown in the figure).

Three kinds of jobs may enter the fabric: grid jobs (from the grid level above), local user jobs, and maintenance jobs (injected by the system itself).

Security Management

Security has always been a high-priority requirement that has been continually addressed by Grid developers. New techniques and technologies are currently being developed to ensure that Grid networks are highly secure. For example, different types of segmentation techniques used for Grid network resources, especially at the physical level, provide capabilities allowing high-security data traffic to be completely isolated from other types of traffic. Also, recently, new techniques using high-performance encryption for Grid networks have been designed to provide enhanced security to levels difficult to obtain on traditional data networks.

5.6 Multimedia Networks

5.6.1 Introduction of Multimedia Networks

Multimedia

Multimedia are media and content that utilizes a combination of different content forms. Multimedia includes a combination of text, audio, still images, animation, video, and interactivity content forms.

Multimedia are usually recorded and played, displayed, or accessed by information content processing devices, such as computerized and electronic devices but can also be part of a live performance. The term "rich media" is synonymous for interactive multimedia. Hypermedia can be considered one particular multimedia application.

Multimedia may be broadly divided into linear and nonlinear categories. Linear active content progresses without any navigation control for the viewer such as a cinema presentation. Nonlinear content offers user interactivity to control progress as used with a computer game or used in self-paced computer based training. Hypermedia are an example of nonlinear content.

Multimedia presentations can be live or recorded. A recorded presentation may allow interactivity via a navigation system. A live multimedia presentation may allow interactivity via an interaction with the presenter or performer.

Major Characteristics of Multimedia

Multimedia presentations may be viewed in person on stage, projected, transmitted, or played locally with a media player. A broadcast may be a live or recorded multimedia presentation. Broadcasts and recordings can be either analog or digital electronic media technology. Digital online multimedia may be downloaded or streamed. Streaming multimedia may be live or on-demand.

Multimedia games and simulations may be used in a physical environment with special effects, with multiple users in an online network, or locally with an offline computer, game system, or simulator.

The various formats of technological or digital multimedia may be intended to enhance the users' experience, for example, to make it easier and faster to convey information. Or in entertainment or art to transcend everyday experience.

Enhanced levels of interactivity are made possible by combining multiple forms of media content. Online multimedia are increasingly becoming object-oriented and data-driven, enabling applications with collaborative end-user innovation and personalization on multiple forms of content over time. Examples of these range from multiple forms of content on Web sites like photo galleries with both images (pictures) and title (text) user-updated, to simulations whose coefficients, events, illustrations, animations, or videos are modifiable, allowing the multimedia "experience" to be altered without reprogramming. In addition to seeing and hearing, Haptic technology enables virtual objects to be felt. Emerging technology involving illusions of taste and smell may also enhance the multimedia experience.

Multimedia Networks

Multimedia networks, currently in their infancy, must support this demanding blend of data. The difficulty of multimedia networking is twofold.

First, the data stream is highly bandwidth-intensive due to the impact of video and image information.

Second, multimedia are isochronous in nature, as the voice and video elements are stream-oriented.

Therefore, the networks that support multimedia must provide substantial bandwidth. They also must accept and deliver the data stream on a regular, continuous, and reliable basis. Such network performance is found only in a few instances, including circuit switched (expensive), dedicated (expensive), or cell-switched (expensive) networks. In case you missed it, expensive is the operative word. While multimedia can be supported at lower cost through ISDN BRI, the bandwidth available is not sufficient to support high-quality video. Additionally, ISDN is relatively expensive (compared to analog circuits) and certainly is not ubiquitous. Multimedia networking in the LAN environment also is problematic. LANs just weren't designed for the task. To be truly effective, multimedia must be networked on an interactive basis.

Typical multimedia network applications include VoIP, IPTV, online game and 3D Internet.

5.6.2 Management of Multimedia Networks

The explosion of IP-based multimedia applications led to an exponential growth of IP traffic and initiated a number of research activities that would allow efficient Quality of Service (QoS) support, such as delay, jitter, loss, and bandwidth allocation.

Network management solutions for these environments must deal with the unique features of multimedia networks: lightweight devices, mobility, connection-oriented operation, Quality of Service guarantees, security, adequate charge-back mechanisms, intelligent filtering/correlation of alarms, and integrated traffic requirements [GSP00].

For fault management:

1) Maintaining and examining ever-increasing error logs.

2) Tracing faults back to devices that have no fixed location in the network (e.g., "personal" routers, devices networked via temporary connections)

3) Multimedia networks tend to generate congestion errors that cannot be considered "faults". The distinction between genuine faults and errors might be more difficult to make.

4) The gathering of faults-related data is more resource intensive. Improved distribution of the processing capabilities and/or storage facilities is needed.

For configuration management:

1) Keeping track of all virtual connections and their re-routing.

2) Monitoring of resources as compared to QoS offered to terminals. In terms of uplink, downlink, and control: being able to change things on the fly.

3) Dual stack configurations and protocol tuning. You might want to configure your device so that it can load a particular stack depending on the type of connection required at the moment (wired vs. wireless). You might want to be able to fine-tune your stack configuration according to the characteristics of the available access point.

4) Configuration of alarm thresholds. Multimedia applications require alarm-setting functions capable of dealing with a more sophisticated mix of traffic on the network.

For performance management:

Guaranteeing the desired service level through configuring connection parameters, setting traffic priorities, and controlling high-volume and high-speed traffic flows becomes an important priority.

1) Monitoring function to spot check QoS delivery satisfying commitments made.

2) Monitoring/reporting of errors on the wireless link, control techniques, forward error correction (FEC), retransmissions, adaptive error control schemes.

3) Production of reports to aid on the planning of additional cache servers, proxy servers, or other high-congestion systems in the network.

4) Monitoring of response times, alarm thresholds, and policies.

5) Monitoring QoS degradation during peak times with the view of finding optimal algorithms.

6) In the case of wireless connections: production of statistics on the air interface, including link usage, connection requests, and connection rejects.

7) Keeping track of network usage by different traffic classes (voice, video, etc.). This information is useful for network planning.

For account management:

Accounting management is needed to help control resource allocation and consumption. Accounting and billing mechanisms that are able to keep track of the constantly changing configuration of the network are required. It is important for an organization to accurately allocate costs to different departments or branches even when many of their users are utilizing the resources of the network in unpredictable and difficult-to-classify patterns.

1) Keeping track of user-based and application-based accounting profile and statistics.

2) Discount or credit schemes for unsatisfied QoS combined with performance management.

3) Management of special tariffs according to the characteristics of the particular connection (service classes).

4) Charging for special routing policies (for example, using predetermined secure links).

For security management:

1) Performing security checks when crossing domains and a looser check when moving in the same domain.

2) Mechanisms to perform power-on authentication.

3) Mechanisms to guarantee privileges on network resource usage.

4) Management of encryption procedures over preselected routes.

5.7   Satellite Networks

5.7.1   Introduction of Satellite Networks

A satellite system or a part of a satellite system, consisting of only one satellite and the cooperating earth stations. A communication satellite is basically an electronic communication package placed in orbit whose prime objective is to initiate or assist communication transmission of information or message from one point to another through space. The information transferred most often corresponds to voice (telephone), to video (television), and to digital data.

Depending on the altitude, satellites can be classified into three types: Low Earth Orbit (LEO), Medium Earth Orbit (MEO), and Geosynchronous Earth Orbit (GEO). GEO satellites are stationary with respect to earth and require fewer ground stations. LEO satellites rotate around the earth at a lower altitude than GEO satellites and require larger number of ground stations. LEO satellites are handed off between ground stations and require mobility management to maintain continuous connectivity with hosts on the ground.

Satellite communications deserve the special merit to allow connecting people at great distances by using the same (homogeneous) communication system and technology. Other very significant advantages of the satellite approach are: (1) easy fruition of both broadcast and multicast high bit-rate multimedia services; (2) provision of backup communication services for users on a global scale (this feature is very important for emergency scenarios and disaster relief activities); (3) provision of services in areas that could not be reached by terrestrial infrastructures; (4) support of high-mobility users.

There are two types of transmission technologies: broadcast and point-to-point transmissions. Satellite networks can support both broadcast and point-to-point connections. Satellite networks are most useful where the properties of broadcast and wide coverage are important. Satellite networking plays an important role in providing global coverage. There are three types of roles that satellites can play in communication networks: access network, transit network, and broadcast network [Sun05].

- Access network

  The access network provides access for user terminals or private networks. Historically, in telephony networks, it provided connections from telephone or private branch exchanges (PBX) to the telephony networks. The user terminals link to the satellite earth terminals to access satellite links directly. Today, in addition to the telephony access network, the access networks can also be the ISDN access, B-ISDN access, and Internet access.

- Transit network

  The transit network provides connection between networks or network switches. It often has a large capacity to support a large number of connections for network traffic. Users do not have direct access to it. Therefore, they are often transparent to users, though they may notice some differences due to propagation delay or quality of the link via a satellite network. Examples of satellite as transit networks include interconnect international telephony networks, ISDN, B-SDN, and Internet backbone networks. Bandwidth sharing is often preplanned using fixed assignment multiple access (FAMA).

- Broadcast network

  Satellite supports both telecommunication service and broadcast service. Satellite can provide very efficient broadcasting services, including digital audio and video broadcast (DVB-S) and DVB with return channels via satellite (DVB-RCS).

Figure 5.24 presents example applications of a satellite network.
Characteristic of Satellite Networks:
Satellite networks have several characteristic features. These include [Kol02]:

- Circuits that traverse essentially the same radio frequency (RF) path length regardless of the terrestrial distance between the terminals.

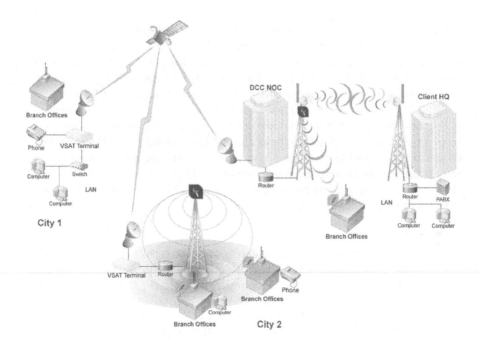

Figure 5.24:   Sample Applications of a Satellite Network

- Circuits positioned in geosynchronous orbits may suffer a transmission delay of about 119 ms between an earth terminal and the satellite, resulting in a user-to-user delay of 238 ms and an echo delay of 476 ms.

- Satellite circuits in a common coverage area pass through a single RF repeater for each satellite link. This ensures that earth terminals, which are positioned at any suitable location within the coverage area, are illuminated by the satellite antenna(s). The terminal equipment could be fixed or mobile on land or mobile on ship and aircraft.

- Although the uplink power level is generally high, the signal strength or power level of the received downlink signal is considerably low because of 1) high signal attenuation due to free-space loss, 2) limited available downlink power, and 3) finite satellite downlink antenna gain, which is dictated by the required coverage area. For these reasons, the earth terminal receivers must be designed to work at significantly low RF signal levels. This leads to the use of the largest antennas possible for a given type of earth terminal and the provision of low-noise amplifiers (LNA) located at close proximity to the antenna feed.

- Messages transmitted via the circuits are to be secured, rendering them inaccessible to unauthorized users of the system. Message security is a commerce closely monitored by the security system designers and users alike.

Satellite communications have several benefits. Satellites provide global coverage and are a good solution for one-way broadcast applications, as one satellite is able to cover a large geographical area. Satellite communications are difficult to disrupt, thus providing a reliable means of communication. Furthermore, mobility is easily supported. The decrease in cost have made it possible even for individuals to take advantage of satellite communications, which is no longer limited to governments, military, and large corporations only.

However, satellite communications do not provide support for all wireless networking needs that have emerged, especially if two-way, high-speed communication with short delays is required.

5.7.2 Management of Satellite Networks

As the adoption of satellite-based communication networks continues to grow, so do the management challenges.

For example, the convergence of voice, video, and data transmission via satellite is driving increased demands on the network to deliver uninterrupted availability, reliability, and security. And, as more mission critical applications become dependent on the network, tolerance for network problems approaches zero.

In particular, maintaining high availability of satellite networks at remote locations presents a number of unique management challenges for operational and IT staff. Communications are often disrupted because of environmental interference, which can require a dispatch of a service technician to the remote site to re-establish connectivity. Likewise, routine network maintenance such as reprovisioning an antenna controller or upgrading a router's operating system with the latest security patch often necessitates a costly on-site visit.

Mobility Management

Mobility management consists of two components: location management and handover management [AMH+99].

- Location management is a two-stage process that enables the network to discover the current attachment point of the mobile user for call delivery. The first stage is location registration (or location update). In this stage, the mobile terminal periodically notifies the network of its new access point, allowing the network to authenticate the user and revise the user's location profile. The second stage is call delivery. Here the network is queried for the user location profile and the current position of the mobile host is found.

Current techniques for location management involve database architecture design and the transmission of signaling messages between various components of a signaling network. As the number of mobile subscribers increases, new or improved schemes are needed to support effectively a continuously increasing subscriber population. Other issues include: security; dynamic database updates; querying delays; terminal paging methods; and paging delays.

- Handoff (or handover) management enables the network to maintain a user's connection as the mobile terminal continues to move and change its access point to the network. There is a three-stage process for handoff. The first stage involves initiation, where either the user, a network agent, or changing network conditions identify the need for handoff. The second stage is new connection generation, where the network must find new resources for the handoff connection and perform any additional routing operations. Under network-controlled handoff (NCHO), or mobile-assisted handoff (MAHO), the network generates a new connection, finding new resources for the handoff, and performing any additional routing operations. For mobile-controlled handoff (MCHO), the mobile terminal finds the new resources and the network approves. The final stage is data-flow control, where the delivery of the data from the old connection path to the new connection path is maintained according to agreed-upon service guarantees.

  Handoff management includes two conditions: intracell handoff and intercell handoff. Intracell handoff occurs when the user moves within a service area (or cell) and experiences signal strength deterioration below a certain threshold that results in the transfer of the user's calls to new radio channels of appropriate strength at the same base station (BS). Intercell handoff occurs when the user moves into an adjacent cell and all of the terminal's connections must be transferred to a new BS. While performing handoff, the terminal may connect to multiple BS's simultaneously and use some form of signaling diversity to combine the multiple signals. This is called soft handoff. On the other hand, if the terminal stays connected to only one BS at a time, clearing the connection with the former BS immediately before or after establishing a connection with the target BS, then the process is referred to as hard handoff.

To allow host mobility, Internet Engineering Task Force (IETF) designed Mobile IP (MIP) and MIPv6. Although MIP or MIPv6 solves the problem of host mobility, it suffers from signaling overhead, handoff latency and inefficient routing.

Host mobility management is not an effective method of managing the mobility of hosts that are moving together, such as in a vehicle, train, or satellite. Use of MIP for managing the mobility of such hosts result in inefficiencies such as increased signaling overhead, increased power consumption, requirement for each host to have powerful transceiver to communicate with the access router, etc. Moreover, nodes that are not capable of running MIP cannot communicate with the outside world.

To solve the problem of aggregate mobility of hosts, IETF has proposed NEtwork MObility (NEMO) where the hosts that move together are connected in a LAN, and the router in the LAN manages the mobility for all the hosts on the LAN. The NEMO Basic Support Protocol (NEMO BSP) from IETF is a logical extension of MIPv6. NEMO BSP performs better than MIPv6 to handle mobility of a large number of nodes, though NEMO BSP has several limitations such as inefficient route for packets, header overhead, and all the limitations inherited from MIPv6 that limits its realization in practical networks.

Resource Management

Satellite resources (i.e., radio spectrum and transmission power) are costly and satellite communications impose special constraints with respect to terrestrial systems in terms of path loss, propagation delay, fading, etc. These are critical factors for supporting user service-level agreements and Quality of Service (QoS).

The ISO/OSI reference model and the Internet protocol suite are based on a layered protocol stack. Protocols are designed such that a higher-layer protocol only makes use of the services provided by the lower layer and is not concerned with the details of how the service is being provided; protocols at the different layers are independently designed. However, there is tight interdependence between layers in IP-based next-generation satellite communication systems. For instance, transport layer protocols need to take into account large propagation delays, link impairments, and bandwidth asymmetry. In addition to this, error correction schemes are implemented at physical, link, and (in some cases) transport layers, thus entailing some inefficiencies and redundancies. Hence, strict modularity and layer independence of the layered protocol model may lead to a nonoptimal performance [Gia07].

Satellite resources must be efficiently utilized in order to provide suitable revenue to operators. Users, however, do not care about the platform technology adopted and employed resource management scheme, but need QoS provision. Unfortunately, resource utilization efficiency and QoS support are conflicting needs: typically, the best utilization is achieved in the presence of a congested system, where QoS can difficulty be guaranteed. A new possible approach addressing both these issues is represented by the cross-layer design of the air interface, where the interdependency of protocols at different layers is exploited with the aim to perform a joint optimization or a dynamic adaptation. The innovation of this approach relies on the fact that it introduces direct interactions event between nonadjacent protocol layers with the aim to improve system performance.

Security Management

Despite its cost-effectiveness over large groups, satellite communication imposes significant security problems. Perhaps the most crucial security issue on satellite networks is the broadcast nature of satellite links. This makes eavesdropping much easier than the fixed terrestrial or mobile networks. Also, data packets over satellite links have high bit-error rates and consequently high packet loss, which might result in loss of security context and security update. Since many of the multicast applications are pay-by, they have to enforce a security mechanism to provide service to only authorized users. Multicast applications are highly dynamic most of the time and there are frequent member changes that require an efficient key update mechanism.

On developing a security framework, especially a key management scheme for satellite communication, one should take the challenges, e.g., limited processing, packet losses, large number of group members, into account.

The security services that have to be provided by satellite networks are stated below [AA06]:

- Confidentiality

  Confidentiality aims not to disclose the sensitive information to unauthorized parties. It protects data from passive attacks, which is called eavesdropping. This service has to be provided particularly in satellite pay-by systems as well as security sensitive distributed applications, where unauthorized users must not access to the service (movie, football match, data, etc.) even if the data are transmitted in broadcast mode. Another aspect of confidentiality is the protection of traffic flow from analysis. Traffic flow protection keeps the attacker away from seeing the source address, destination address, frequency, and other characteristics of the packet.

- Authentication and Nonrepudiation

  Authentication aims to prove the identity of an entity as it claims to be. It is necessary to have a strong authentication mechanism since impersonation is easier on the wireless channels. There are three specific authentication services defined:

  Peer entity authentication: The authentication is established at times during the data transfer phase. It attempts to make sure that the connection is not replayed. This is mostly provided through handshake messages which include nonces and/or timestamps.

  Data origin authentication: The data source of a connection is authenticated. This service does not protect against duplication or modification of data units.

  Group authentication: This service ensures that the received multicast messages by group members originate from a valid group member (regardless of its identity).

- Integrity

  This service ensures that the message is received as sent. It provides protection against modification of packets. In data security context, cryptographic integrity could be used to provide authenticated data to end terminals. It is an application specific issue to consider whether the system has to provide data origin, group or even peer-entity authentication and is not explored within the context of this paper.

- Key Management

  Key is the essential component of the security of most cryptographic operations. Thus, maintaining key management is an important task.

  Providing all the above-mentioned security services depends basically on an efficient and secure key management. Even if the cryptographic primitives are secure, the system is vulnerable without a secure key management protocol.

  A key management protocol is the set of processes and mechanisms to support establishment and maintenance of keys, throughout the life cycle of keying material. There are different solutions for key management in group communication. One approach is to use a central server to distribute the secret decryption key (group key) to the legitimate receivers. Another approach is to use a contributory technique where the key is established with the contribution of members. A famous contributory group key management technique where there is no central server, is Group Diffie-Hellman (GDH).

5.8 Storage Networks

5.8.1 Introduction of Storage Network

Storage network refer to a group of networked, externally attached storage devices, together with their associated control and switching components, that are managed as a single entity. Storage network can be classified as Direct Attached Storage (DAS), Storage Area Networks (SAN), and Network Attached Storage (NAS) [Jep03].

- Direct Attached Storage (DAS) is the term used to describe a storage device that is directly attached to a host system. The simplest example of DAS is the internal hard drive of a server computer, though storage devices housed in an external box come under this banner as well. DAS is still, by far, the most common method of storing data for computer systems. Over the years, though, new technologies have emerged that work, if you'll excuse the pun, out of the box.

- Network Attached Storage (NAS) is a data storage mechanism that uses special devices connected directly to the network media. These devices are assigned an IP address and can then be accessed by clients via a server that acts as a gateway to the data or, in some cases, allows the device to be accessed directly by the clients without an intermediary.

The beauty of the NAS structure is that it means that in an environment with many servers running different operating systems, storage of data can be centralized, as can the security, management, and backup of the data. An increasing number of companies already make use of NAS technology, if only with devices such as CD-ROM towers (stand-alone boxes that contain multiple CD-ROM drives) that are connected directly to the network.

Some of the big advantages of NAS include the expandability; need more storage space, add another NAS device, and expand the available storage. NAS also bring an extra level of fault tolerance to the network. In a DAS environment, a server going down means that the data that that server holds is no longer available. With NAS, the data are still available on the network and accessible by clients.

- Storage Area Network (SAN) is a network of storage devices that are connected to each other and to a server, or cluster of servers, which act as an access point to the SAN. In some configurations a SAN is also connected to the network. SAN's use special switches as a mechanism to connect the devices. These switches, which look a lot like a normal Ethernet networking switch, act as the connectivity point for SAN's. Making it possible for devices to communicate with each other on a separate network brings with it many advantages. Consider, for instance, the ability to back up every piece of data on your network without having to "pollute" the standard network infrastructure with gigabytes of data.

Historically, data centers first created "islands" of SCSI disk arrays. Each island was dedicated to an application, and visible as a number of "virtual hard drives" (i.e., LUNs). Essentially, a SAN connects storage islands together using a high-speed network, thus allowing all applications to access all disks.

Operating systems still view a SAN as a collection of LUNs, and usually maintain their own file systems on them. These local file systems, which cannot be shared among multiple operating systems/hosts, are the most reliable and most widely used. If two independent local file systems resided on a shared LUN, they would be unaware of this fact, would have no means of cache synchronization, and eventually would corrupt each other. Thus, sharing data between computers through a SAN requires advanced solutions, such as SAN file systems or clustered computing. Despite such issues, SANs help to increase storage capacity utilization since multiple servers share the storage space on the disk arrays. The common application of a SAN is for the use of transactionally accessed data that require high-speed, block-level access to the hard drives such as e-mail servers, databases, and high-usage file servers.

In contrast, NAS allows many computers to access the same file system over the network and synchronizes their accesses. Lately, the introduction of NAS heads allowed easy conversion of SAN storage to NAS. See Figure 5.25 and Figure 5.26.

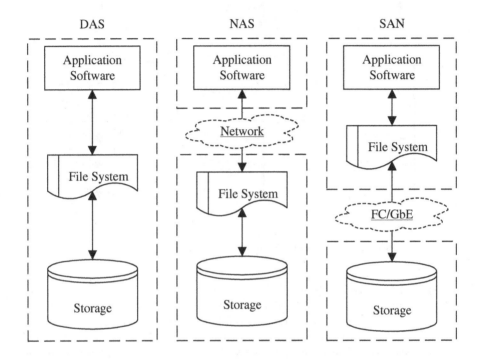

Figure 5.25: SAN vs. NAS vs. DAS

Despite the differences between NAS and SAN, it is possible to create solutions that include both technologies, as shown in the diagram (see Figure 5.27).

Sharing storage usually simplifies storage administration and adds flexibility since cables and storage devices do not have to be physically moved to shift storage from one server to another.

Other benefits include the ability to allow servers to boot from the SAN itself. This allows for a quick and easy replacement of faulty servers since the SAN can be reconfigured so that a replacement server can use the LUN of the faulty server. This process can take as little as half an hour and is a relatively new idea being pioneered in newer data centers. There are a number of emerging products designed to facilitate and speed this up still further. Brocade, for example, offers an Application Resource Manager product which automatically provisions servers to boot off a SAN, with typical caseload times measured in minutes. While this area of technology is still new, many view it

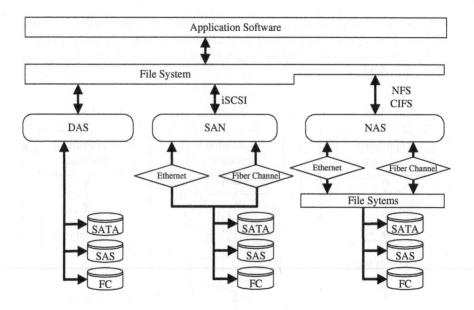

Figure 5.26: Organization of SAN, NAS, and DAS

as being the future of the enterprise data center.

SANs also tend to enable more effective disaster recovery processes. A SAN could span a distant location containing a secondary storage array. This enables storage replication either implemented by disk array controllers, by server software, or by specialized SAN devices. Since IP WANs are often the least costly method of long-distance transport, the Fiber Channel over IP (FCIP) and iSCSI protocols have been developed to allow SAN extension over IP networks. The traditional physical SCSI layer could only support a few meters of distance, not nearly enough to ensure business continuance in a disaster.

The economic consolidation of disk arrays has accelerated the advancement of several features, including I/O caching, snapshotting, and volume cloning (Business Continuance Volumes or BCVs).

5.8.2 Management of Storage Networks

The management of storage networks is of different significance to various technical fields. For example, the classical network administrator is interested in how the data should be transported and how it is possible to ensure that the transport functions correctly. Further aspects are the transmission capacity of the transport medium, redundancy of the data paths, or the support for and operation of numerous protocols (Fiber Channel FCP, iSCSI, NFS, CIFS, etc.). To a network administrator, it is important how the data

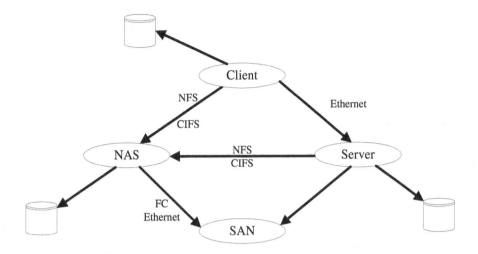

Figure 5.27: Storage Network with NAS and SAN

travels between the hosts and not what happens to it when it finally arrives at its destination.

However, for a storage network management, the administrator should consider the organization and storage of the data when it has arrived at its destination.

A balanced management system must ultimately live up to all these different requirements equally. It should cover the complete bandwidth from the start of the conceptual phase through the implementation of the storage network to its daily operation. Therefore, right from the conception of the storage network, appropriate measures should be put in place to subsequently make management easier in daily operation.

A good way of taking into account all aspects of such a management system for a storage network is to orient ourselves with the requirements that the individual components of the storage network will impose upon a management system. These components include [TEM04]:

- Applications

  These include all software that processes data in a storage network.

- Data

  Data is the term used for all information that is processed by the applications, transported over the network, and stored on storage resources.

- Resources

  The resources include all the hardware that is required for the storage and the transport of the data and the operation of applications.

- Network

  The term network is used to mean the connections between the individual resources. Diverse requirements can now be formulated for these individual components with regard to monitoring, availability, performance, or scalability. Some of these are requirements such as monitoring that occur during the daily operation of a storage network, others are requirements such as availability that must be taken into account as early as the implementation phase of a storage network. For reasons of readability we do not want to investigate the individual requirements in more detail at this point. In Appendix B you will find a detailed elaboration of these requirements in the form of a checklist. We now wish to turn our attention to the possibilities that a management system can offer in daily operation.

Generally, the following management disciplines are all part of Enterprise Storage Resource Management:

- Asset management

  This discipline addresses the need to discover the resources, recognize the resource, and tie it to the rest of the topology. This means that an agent could distinguish the difference between a high-end DASD (Direct Access Storage Disk) storage facility, a Fiber Channel switch, high-end virtual tape server, or other resource. After discovery, it would dynamically load the latest version of an agent and call an API for asset information. It would probably contain information such as:

  There are many functions that could be put under this discipline such as asset discovery, asset topology, asset lease management, and software and microcode management.

- Capacity management

  This set of information would vary depending on the resource being managed. For example, in large DASD storage facilities, we would need to understand multiple levels of capacity. Basically, IT departments don't ever want to run out of free space.

  To do positive capacity planning, corporate resource managers need to understand the additional capacity available at both the physical and the logical storage levels. This includes information like a box's available free space/slots, unassigned volumes, free/used space within the assigned volumes, plus some file-level detail. They also need to understand the growth capacity based on the model, or how many frames with slots for disk drawers could be added if necessary. For a Fiber Channel switch, the capacity could be expressed as a data transfer rate based on the horsepower of the device or the number of ports.

In software, necessary information might include the number of backups, backup tapes, percent utilization, and percent scratch. IT management needs answers to question such as "If I backup this application, what will it do to my network and how much back-end storage will I need to hold it?"

In mainframe environments, this technology is a mature science. In a world of open systems connected to a set of high-end, multi-platform storage facilities, it is embryonic. VTOCs (Volume Table Of Contents), and VVDSs (VSAM VTOC Data Set), provide many of the answers on S/390 platforms. In order to provide the function for the enterprise, the ESRM software must understand every flavor of operating platform, all of the platform file systems, the configuration of every vendor device and their associated interfaces, and every flavor of storage management software.

- Configuration management

Initially, it might seem virtually impossible to think about a common API, which would allow ESRM (Enterprise Storage Resource Management) software to configure all OEM (Original Equipment Manufacturers) storage facilities. But on further investigation, there are many similarities in all storage facilities devices. Today, much of this information is kept at the host level. To allow an ESRM agent to collect this data directly, the storage facility would need to keep track of all performance counters.

Consider the high-end DASD systems being built today. They all have cache, some with multiple layers like drawer cache. Some have NVS (nonvolatile storage) for writes; others emulate this in the read cache. All have host adapters in the upper interfaces of the device. Each host adapter has a certain number of ports of various flavors, which connect to the host. All vendors have lower interfaces to the disks, called disk adapters, which have a number of ports that connect to various transport types (SSA, SCSI, etc.). All have DDMs (Disk Device Modules), which usually fit into drawers and have varying amounts of raw storage capacity. There are only so many RAID types supported by vendors that can be addressed by a common API.

Other notions, such as sparing, mirroring, remote copy, and instantaneous copy also have common threads that could be represented in industry-wide models. All storage facilities have the notion of logical configuration and physical configuration data, and the means to switch between them easily.

Fiber Channel configurations, have topology connectivity management for hosts, devices, and the interconnect fabric of hubs and switches. Host connectivity is through adapters, which have a certain number of ports of a certain type. Hosts feed into switches in the fabric, which can either connect directly to storage adapters, to other switches, or hubs. Hubs can be cascaded to other hubs or to device adapters.

IT departments need to be able to see the current configuration and they need to understand when a physical failure occurs, and what application(s) were affected. They need to be able to set the configuration based on business requirements such as high availability and high accessibility. Lastly, they need to be able to do this to any OEM device through the same user interface.

- Performance management

  In a world of high-end, multi-platform, intelligent subsystems, it becomes critical to do more than the standard, classical performance analysis of problem isolation with the upper or lower interfaces, cache, and NVS overload. IT managers must drill-down to the top volumes, and determine the platform, the application, and even the file causing the problem. Today this is impossible because there are no common platform-independent APIs to access standard, reliable performance information from all OEM storage facilities.

  For Fiber Channel management, it may involve the management of zoning to ensure that the critical business applications get the bulk of the traffic capacity. It may also include the recognizing of trapped packets, which are stuck in the fabric but eating up the latent capacity.

  Performance management of virtual tape servers might include things like monitoring the DASD buffer for hit-ratios of virtual tapes, and virtual mounts very similar to cache management in DASD storage facilities.

  Management software would include the ability to monitor automatic workloads such as backup or space management compared to the window they are expected to run in so that alerts can be externalized to start additional workload tasks if necessary.

- Availability management

  Availability management is about the prevention of failure, correction of problems as they happen, and the warning of key events long before the situation becomes critical.

  For example, monitoring of the number of I/O errors on a tape head to automatically mount a cleaning cartridge is a good example of availability management. Another example would be a high-availability function that upon the failure of a DASD mirrored pair would search for a spare, break the mirrored pair, re-mirror the good drive with the spare, and page the customer engineer to repair the bad drive so that the system does not go down. Indeed, one common thread in all data centers today is that there are fewer people to manage the ever-growing farm of enterprise storage. Reports, graphs, and real-time monitoring are useful, but only to a point. There are no people to sit in front of "GUI Glow Meters" to monitor the system. ESRM software must provide easy automation trigger events tied-in with policies and thresholds to allow the monitoring function to operate without people. There is an infinite set of automation and policy management functions that could be provided under ESRM software.

For example, if IBM DFSMShsm is half-way through the Primary Space Management window but only one-quarter of the way through the volumes, then there is a good chance that it won't complete. If PSM does not complete, then this company won't have the available free space to do their business every day. The real-time monitoring would let the storage administrator see this as it was happening. A report or graph will let the storage administrator know this after the fact.

- Outboard management

  This discipline addresses the management of hardware that contains built-in data movement and high-data availability functions. There are a lot of useful, time/people-saving functions that could be provided by ESRM software.

  Today, there are many data movement functions being provided by various storage vendors, especially in the high-end DASD storage facilities. The data mining industry and the Y2K problem have created a huge market for data replication products such as DataReach, HDME, TimeFinder, ESP, InfoMover, InfoSpeed, FileSpeed, and SnapShot. The business continuance industry and the disaster/recovery requirements have forced outboard storage technologies for remote data copy with such functions as concurrent copy, PPRC, XRC, and SRDF.

  Although these functions are powerful, they do require some user management not only for the data identification but also the scheduling, start, stop, and error handling. Plus, the user is expected to understand the nuances of every vendor's twist on the particular data/device movement function.

- Policy management

  Policy management is probably the most nebulous ESRM discipline. The scope of policies has such a large range of possibilities. For example, imagine a simple policy, which states that if any port of a Fiber Channel switch goes down, then the appropriate person should be paged. This is fairly straightforward. As we move up the food chain in this discipline, we see more complex possibilities. How about a policy that states that you never want to run out of free space? Or, how about specifying an average of 6 milliseconds or less on every I/O against a file that has PROD as the second-level qualifier?

There may not be single policies that will cover all of the ESRM disciplines. It is clear that users do want the system to manage itself as much as possible and they do want to concentrate on doing whatever is necessary to have a successful business. At a minimum, ESRM software will have to provide primitives to allow automation of basically anything that could (and should) be automated.

Combinations of the primitive policies may form the actual business policies. ESRM should provide the framework for establishing those policies, for setting the controls/thresholds/auto-scripts, and for managing the storage resources based on those policy definitions, thresholds, and controls.

The information that one gets today through channel interfaces from those same high-end, multi-platform, intelligent storage facilities is wonderful. The engineers of all storage manufacturers really bent over backward to preserve the concept of 3880/3990 performance model to allow incredibly accurate classical performance analysis for cache, upper interfaces, lower interfaces, and volumes, when connected to a mainframe only environment. It is needed to extend this wonderful information to the open systems world through an IP connection and to allow easier management of storage resources by not only providing the API but also by specifying a recommended underlying architecture and standards. To support large SAN configurations with all the high availability, reliability, and security expected by today's Enterprise IT centers will require considerable cooperation and coordination among vendors in the storage industry.

5.9   Cognitive Networks

5.9.1   Introduction of Cognitive Networks

Cognitive Computing

Cognition is defined as the mental process or faculty of knowing, including aspects such as awareness, perception, reasoning, and judgment:

- Awareness: Observe the environment

- Perception: Learn from the environment and understand the changes of environment

- Reasoning: Analysis the reasons or the motivations of the changes

- Judgment: Decide what to do according to the results of reasoning to achieve a predefined goal.

Cognitive computing refers to the development of computer systems modeled after the human brain. Cognitive computing integrates technology and biology in an attempt to re-engineer the brain, one of the most efficient and effective computers on earth.

Deeper biological understanding of how the brain worked allowed scientists to build computer systems modeled after the mind and, most important, to build a computer that could integrate past experiences into its system. Cognitive computing was reborn, with researchers at the turn of the 21st century developing computers that operated at a higher rate of speed than the human brain did.

Cognitive computing integrates the idea of a neural network, a series of events and experiences that the computer organizes to make decisions.

For an application model of cognitive computing in distributed networks, by making use of the computing resources available on the nodes, low-level data will be integrated and fused into a smaller amount of information, thus greatly reducing the amount of communications. This will be accomplished by having several expert systems analyze at different levels of abstraction and integration. Intra-nodal processing will include the integration and fusing of complex data from the node's heterogeneous sensors. Internode processing will integrate and fuse nodal information from geographically separated sensors. Finally, by integrating High Performance Computing and sensor networks, phenomena, based models will be used as complex events that trigger the reporting of sensor observations.

Cognitive Radio

Cognitive radio is a paradigm for wireless communication in which either a network or a wireless node changes its transmission or reception parameters to communicate efficiently avoiding interference with licensed or unlicensed users. This alteration of parameters is based on the active monitoring of several factors in the external and internal radio environment, such as radio frequency spectrum, user behavior, and network state. Cognitive radio is also sometimes called smart radio, frequency agile radio, police radio, or adaptive software radio, and so on.

The main functions of Cognitive Radios are as follows [ALV+06]:

- Spectrum Sensing: detecting the unused spectrum and sharing it without harmful interference with other users, it is an important requirement of the Cognitive Radio network to sense spectrum holes, detecting primary users is the most efficient way to detect spectrum holes. Spectrum sensing techniques can be classified into three categories:

  - Transmitter detection: cognitive radios must have the capability to determine whether a signal from a primary transmitter is locally present in a certain spectrum, there are several approaches proposed:
    * matched filter detection
    * energy detection

           * cyclostationary feature detection

     – Cooperative detection: refers to spectrum sensing methods where information from multiple Cognitive radio users are incorporated for primary user detection.

     – Interference-based detection.

- Spectrum Management: Capturing the best available spectrum to meet user communication requirements. Cognitive radios should decide on the best spectrum band to meet the Quality of service requirements over all available spectrum bands; therefore, spectrum management functions are required for Cognitive radios. These management functions can be classified as:

     – spectrum analysis

     – spectrum decision

- Spectrum Mobility: Defined as the process when a cognitive radio user exchanges its frequency of operation. Cognitive radio networks target to use the spectrum in a dynamic manner by allowing the radio terminals to operate in the best available frequency band, maintaining seamless communication requirements during the transition to better spectrum

- Spectrum Sharing: Providing the fair spectrum scheduling method, one of the major challenges in open spectrum usage is the spectrum sharing. It can be regarded to be similar to generic media access control MAC problems in existing systems

Evolution of Cognitive Radio toward Cognitive Networks is under process, in which Cognitive Wireless Mesh Network (e.g., CogMesh) is considered as one of the enabling candidates aiming at realizing this paradigm change.

Cognitive Networks

A cognitive network is a network composed of elements that, through learning and reasoning, dynamically adapt to varying network conditions in order to optimize end-to-end performance. In a cognitive network, decisions are made to meet the requirements of the network as a whole, rather than the individual network components. Cognitive networks can be characterized by their ability to perform their tasks in an autonomous fashion by using their self-attributes such as self-managing, self-optimizing, self-monitoring, self-repair, self-protection, self-adaptation, self-healing to adapt dynamically to changing requirements or component failures while taking into account the end-to-end goals.

Cognitive networks use self-configuration capability to respond and dynamically adapt to the operational and context changes. Main function components of self-configuration are self-awareness and auto-learning that are implemented by means of network-aware middleware and normally distributed across the network components. Applications and devices adapt to exploit enhanced network performance and are agnostic of the underlying reconfigurations, in accordance with the seamless service provision paradigm.

Cognitive wireless access networks are those that can dynamically alter their topology and/or operational parameters to respond to the needs of particular user while enforcing operating and regulatory policies and optimizing overall network performance. A cognitive infrastructure consists of reconfigurable elements and intelligent management functionality that will progressively evolve the policies based on the past actions.

Figure 5.28 illustrate cogitative radio network architecture [ALV+08].

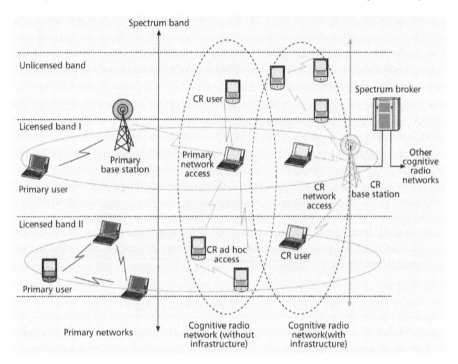

Figure 5.28: Cognitive Radio Network Architecture

The realization of cognitive, wireless access networks requires intelligent management functionality, which will be in charge of finding the best reconfigurations. See Figure 5.29.

Wireless solutions based on cognitive network principles encompass technologies and products to ensure that the networks, network components, as well as networked devices and applications, can be deployed and managed (configured, optimized, healed, and protected), in realtime. Cognitive networks feature a distributed management functionality that can be implemented in accordance with the autonomic computing paradigm. The holistic,

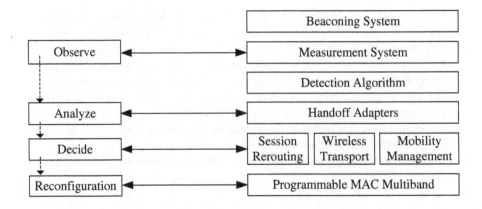

Figure 5.29: An Example of Cognitive Network Functionality

collective cooperation and action of the distributed, autonomic components yields a self-healing and scalable solution that accounts for the potential evolution of services and growing user needs.

Wireless solutions based on cognitive network principles remove shortfalls of cooperative networks such as interoperator dependencies, frequent infrastructure upgrades and challenges of split network management.

At the same time, cognitive networks maximize the operator's ability to:

- Benefit from economies of scale introduced by common hardware platforms and software architectures supporting evolution of radio access solutions

- Improve time-to-market performance by supporting new service offerings without the need to upgrade the infrastructure

- Maximize return-on-investment by maximizing the exploitation of available/deployed resources

- Accelerate innovation by enabling opportunistic usage of spectrum resource, dynamically adjusting its Tx and Rx parameters to exploit unused spectrum at any given location at any point in time

Cognitive networks are important because they are capable of rendering efficient, ubiquitous, pervasive (ambient), and context-aware application provision. This can be referred to as a consistent seamless mobility experience that bridge across the connectivity and content/application delivery solution domains. Examples of potential cognitive network applications include [TDM05]:

- Heterogeneity. For networks that employ a wide variety of protocols and physical layer interfaces, a cognitive network can provide a mechanism for creating order in chaos. Since a cognitive network views and learns from the observed network status, it can de-conflict individual nodes and optimize the connections, from top-level objectives such as creating efficient homogeneous clusters to lower-level goals such as reducing the total amount of energy expended.

- QoS. In a more general sense, cognitive networks can be used to manage the QoS for a connection. Utilizing the feedback about observed network conditions, the cognitive network can identify bottlenecks, estimate guarantees, change prioritization, and optimize behaviors to provide the desired end-to-end QoS.

- Security. Cognitive networks could also be used for security purposes such as access control, tunneling, trust management, or intrusion detection. By analyzing feedback from the various layers of the network, a cognitive network can find patterns and risks and then react by changing such security mechanisms as rule sets, protocols, encryption, and group membership. Such cognition may also aid in trust and reputation mechanisms.

5.9.2 Management of Cognitive Networks

Cognitive wireless networks are capable of reconfiguring their infrastructure, based upon experience, in order to adapt to continuously changing network environments. Cognitive networks are seen as a major facilitator of future heterogeneous internetworking and management, capable of continuously adapting to fluid network characteristics as well as application-layer QoS requirements.

The main tasks of management of cognitive networks include spectrum management, location management, handoff management, and cross-layer design, etc.

Spectrum Management

Cognitive radio (CR) is a key technology for alleviating the inefficient spectrum utilization problem under the current static spectrum-allocation policy. In cognitive networks (CRNs), unlicensed or secondary users (SUs) are allowed to opportunistically utilize spectrum bands assigned to licensed or primary users (PUs) as long as they do not cause any harmful interference to PUs. A spectrum opportunity refers to a time duration on a channel during which the channel can be used by SUs without interfering with the channel's PUs. In-band channels refer to those channels currently in use by SUs; all others are referred to as out-of-band channels. To efficiently manage and organize spectrum holes information among cognitive radios, good spectrum management scheme is necessary.

CR networks impose unique challenges due to their coexistence with primary networks as well as diverse QoS requirements. Thus, new spectrum management functions are required for CR networks with the following critical design challenges:

- Interference avoidance: CR networks should avoid interference with primary networks.

- QoS awareness: To decide on an appropriate spectrum band, CR networks should support QoS-aware communication, considering the dynamic and heterogeneous spectrum environment.

- Seamless communication: CR networks should provide seamless communication regardless of the appearance of primary users.

The spectrum management process consists of four major steps [ALV+08]:

- Spectrum sensing: A CR user can allocate only an unused portion of the spectrum. Therefore, a CR user should monitor the available spectrum bands, capture their information, and then detect spectrum holes.

- Spectrum decision: Based on the spectrum availability, CR users can allocate a channel. This allocation not only depends on spectrum availability but is also determined based on internal (and possibly external) policies.

- Spectrum sharing: Because there may be multiple CR users trying to access the spectrum, CR network access should be coordinated to prevent multiple users colliding in overlapping portions of the spectrum.

- Spectrum mobility: CR users are regarded as visitors to the spectrum. Hence, if the specific portion of the spectrum in use is required by a primary user, the communication must be continued in another vacant portion of the spectrum.

Figure 5.30 illustrates the spectrum management framework for cognitive networks [ALV+06]. It is evident from the significant number of interactions that the spectrum management functions require a cross-layer design approach.

Location Management

Because of the heterogeneity of CNs, routing and topology information is more and more complex. Good mobility and connection management can help neighborhood discovery, detect available Internet access and support vertical handoffs, which help cognitive radios to select route and networks.

Location management is a two-stage process that enables the network to discover the current attachment point of the mobile user for call delivery

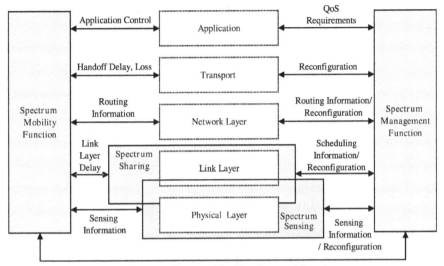

Figure 5.30: Spectrum Management Framework for Cognitive Networks

[CPP+08]. The two stages are location registration and call delivery. When an MS (Mobile Station) visits to a foreign network (FN) and wants to get the Internet access service, it will first discover the mobile agents of the FN by detection agent advertisements. After getting the agent advertisement, the MS is able to form a CoA and inform the HA (Home Agent) the association between the current CoA (Care-of-Address) and MN (Mobile Node)'s home address. However, in CNs, a CR-MS (Cognitive Radio Mobile Station) can simultaneously connect with many different wireless systems, which may belong to different FNs, and it should acquire a CoA from each of them in order to route packets to/from them. So there is a need to develop new schemes to deal with multiple CoAs. Specifically, a CR-MS can acquire many CoAs from each of those connectable FNs so that a CR-MS may no longer use a single CoA to represent its current position and to route packets. Moreover, multiple CR-MSs can form an ad-hoc network and some of them may connect to BSs(Base Stations)/APs(Access Points) and access backbone/core networks. We call these nodes as "gateway nodes." Because of the limited coverage of BSs/APs, some of the MSs can only get the BSs' service through multi-hop relay MSs. The cooperation and integration of Mobile IP and ad-hoc network is a research challenge.

Figure 5.31[1] [AC07] illustrates a model for the location information management system in cognitive networks. According to this model, the measurement and/or sensing devices are used to obtain data from the operational en-

---

[1]With kind permission of Springer Science and Business Media.

vironments. The acquired data are sent to location information management system for post-processing, which is embedded in central cognitive engine of the network and/or the cognitive engine of cognitive radio node. Location estimation and/or sensing algorithms process the data to determine location information. Since different location estimation and/or sensing methods provide the estimated and/or sensed location information in different coordinate systems, cognitive engine needs a coordinate system converter to manage the transition between different coordinate systems. Finally, location information management system utilizes location information for different applications such as location-based services (i.e., positioning and tracking), network optimization, transceiver algorithm optimization, and environment characterization. Furthermore, location information management system has a mechanism to handle mobility and tracking tasks.

Handoff Management

Handoff management enables the network to maintain a user's connection as the mobile terminal continues to move and change its access point to the network. Three stages are included: initiation, new connection generation, and data flow control. Because of the multi-hop characteristic of CN, handover management is no longer an issue between a single MS and FNs. It is about multiple MSs and FNs. For example, if some of the gateway nodes move away from BSs' coverage area, they shall inform those nodes in the ad-hoc network about its loss of connection. So nodes in the ad-hoc network can prepare to perform handover if they have active connections through that gateway node. Moreover, some FNs can never be connectable due to the lost of gateway nodes. This can induce those CoAs issued from those FN to be invalid. New CoA registration mechanisms may also be necessary.

When a CR node turns on, there would be a lot of available access networks around it. Here, available access networks are those networks it is able to get authorized to use their network resources. To be authorized to get network resources and then services, the mobile user should first be trusted. When the user roams around different access networks, the trust relationship should be set up first.

Since a cognitive radio has the reconfigurable ability, any inappropriate modification or adjustment of system parameters should be prohibited. Also, if some nodes in the CN behave maliciously or selfishly, they shall be excluded from the network.

Cross-Layer Design and Optimization

Computer networks have been designed following the principle of protocol layering, so network functionalities are designed in isolation of each other (separate layers) and interfaces between layers. Each layer uses the services provided by the layer below it and provides services to the layer above it.

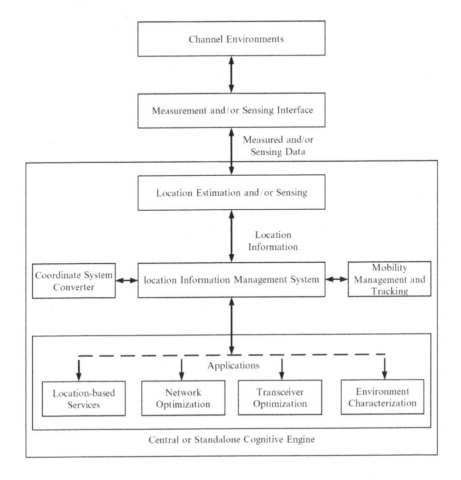

Figure 5.31:   A Model for Location Information Management in Cognitive Networks

Interlayer communication happens only between adjacent layers and is limited to procedure calls and responses, examples of this are the seven-layer Open Systems Interconnection (OSI) and the four-layer TCP/IP model.

In cognitive networks, however, there is a need for greater interaction between the different layers of the protocol stack in order to achieve the end-to-end goals and performance in terms of resource management, security, QoS, or other network goals. Cross-layer design refers to protocol design done by actively exploiting the dependence between the protocol layers to obtain performance gains. Cognitive networks will employ cross-layer design and optimization techniques in order to adapt, simply because a great level of coordination is needed between the traditional protocol layers. Indeed, an efficient protocol stack that responds to the environment, network conditions, and user demands is central to the very idea of cognitive networks. The pursuit of achieving such efficiency and flexibility in wireless networks is not feasible by maintaining the strict boundaries between the different network functionalities, as done by layering. The way forward is inevitably cross-layer design.

Alongside the cross-layer design proposals, initial proposals on how cross-layer interactions can be implemented by (1) direct communication between layers, (2) a shared database across the layers, and (3) completely new abstractions.

5.10 Future Internet

5.10.1 Introduction of the Internet

The Internet is a global network of computers that communicate via TCP/IP protocol. It is a "network of networks" that consists of millions of private and public, academic, business, and government networks of local to global scope that are linked by copper wires, fiber-optic cables, wireless connections, and other technologies.

The Internet is the basis for the World Wide Web, E-mail, P2P applications, VOIP, and hundreds of other uses. Although the basic data being transmitted in each of these cases may be virtually identical, each case requires special treatment of the data in regard to data integrity, speed, redundancy, and error correction. The success of the Internet is therefore attributed to its flexibility in providing a platform for the differing data protocols and their individual needs.

Aside from the complex physical connections that make up its infrastructure, the Internet is facilitated by bi- or multi-lateral commercial contracts (e.g., peering agreements), and by technical specifications or protocols that describe how to exchange data over the network. Indeed, the Internet is defined by its interconnections and routing policies.

The complex communications infrastructure of the Internet consists of its hardware components and a system of software layers that control various aspects of the architecture. While the hardware can often be used to support other software systems, it is the design and the rigorous standardization process of the software architecture that characterizes the Internet.

The most prominent component of the Internet model is the Internet Protocol (IP), which provides addressing systems for computers on the Internet and facilitates the internetworking of networks. IP Version 4 (IPv4) is the initial version used on the first generation of the today's Internet and is still in dominant use. It was designed to address up to about 4.3 billion (10^9) Internet hosts. However, the explosive growth of the Internet has led to IPv4 address exhaustion. A new protocol version, IPv6, was developed which provides vastly larger addressing capabilities and more efficient routing of data traffic. IPv6 is currently in commercial deployment phase around the world.

IPv6 is not interoperable with IPv4. It essentially establishes a "parallel" version of the Internet not accessible with IPv4 software. This means software upgrades are necessary for every networking device that needs to communicate on the IPv6 Internet. Most modern computer operating systems are already converted to operate with both versions of the Internet Protocol. Network infrastructures, however, are still lagging in this development.

There have been many analyses of the Internet and its structure. For example, it has been determined that the Internet IP routing structure and hypertext links of the World Wide Web are examples of scale-free networks.

The Internet Corporation for Assigned Names and Numbers (ICANN) is the authority that coordinates the assignment of unique identifiers on the Internet, including domain names, Internet Protocol (IP) addresses, and protocol port and parameter numbers. A globally unified namespace (i.e., a system of names in which there is at most one holder for each possible name) is essential for the Internet to function. ICANN's role in coordinating the assignment of unique identifiers distinguishes it as perhaps the only central coordinating body on the global Internet, but the scope of its authority extends only to the Internet's systems of domain names, IP addresses, protocol ports, and parameter numbers.

The Internet is allowing greater flexibility in working hours and location, especially with the spread of unmetered high-speed connections and Web applications.

The Internet can now be accessed virtually anywhere by numerous means. Mobile phones, datacards, handheld game consoles, and cellular routers allow users to connect to the Internet from anywhere there is a cellular network supporting that device's technology.

Many computer scientists see the Internet as a "prime example of a large-scale, highly engineered, yet highly complex system." The Internet is extremely heterogeneous. (For instance, data transfer rates and physical characteristics of connections vary widely.) The Internet exhibits "emergent phenomena" that depend on its large-scale organization. For example, data transfer rates exhibit temporal self-similarity. Further adding to the complexity of the Internet is the ability of more than one computer to use the Internet through only one node, thus creating the possibility for a very deep and hierarchal subnetwork that can theoretically be extended infinitely (disregarding the programmatic limitations of the IPv4 protocol). However, since principles of this architecture date back to the 1960s, it might not be a solution best suited to modern needs, and thus the possibility of developing alternative structures is currently being looked into.

5.10.2 Future Internet

Evolution of Internet

The Internet is evolving from the interconnection of physical networks by a collection of protocols toward what is considered the future Internet: a network of applications, information, and content. Some characteristics of the future Internet [TTA08]:

- The future Internet can be seen as a network of applications.

- It will enable "peer productivity" and becomes an "architecture for participation".

- In particular, the future Internet will be based on interactive, edge-based applications and overlays, such as peer-to-peer (P2P) content distribution, Skype, MySpace, or YouTube.

- However, it is not yet sure what the next major application in the future Internet is.

The future Internet is no longer a collection of links, routers, and protocols. It will be viewed as a network of applications, information, and contents. The future Internet will become an architecture for participation by the users and, eventually, for contribution of hardware resources. Hence, intelligent edge-based applications and services will dominate the future Internet. These applications and services will be typically implemented in an abstract way as overlays.

Recent advances in networking technology such as high-speed optical networking, wireless transmission, or virtualization of links and routers will challenge the design of the future Internet. In order to address these challenges new methodologies for implementing and operating overlays are needed. In particular new mechanisms are required which permit edge-based overlays to structure their topology, to define their routing scheme, and to manage their resources independently.

Moreover, the pressures from the efficiencies of overlays on the conventional layering model of IP and OSI initiate currently a rethinking of these models. A thinning of protocol layers and a more basic separation of the layers appear essential. This separation should focus on a split into three layers: (1) the application layer (for addressing the application needs), (2) the mediation layer (for network structuring, naming, and routing), and (3) the transport layer (for reliable and cost-efficient transport).

Future Internet Properties

A number of basic properties can already be found in today's Internet and are to be carried forward as they have proven their effective support of the basic requirements found in common usage. These properties include forwarding, routing, encapsulation, tunneling, and so forth. In the following, mainly additional properties are addressed, although some overlap with today's function can also be observed [Pap09].

Considering the multifaceted requirements facing the Future Internet (FI), individual demands should be fulfilled enjoying the scale and scope effects following from a common network. Although without implying prioritization, the functional properties of the FI shall include:

- Accountability

- Security

- Privacy

- Availability (maintainability and reliability)

- Manageability and diagnosability (root cause detection and analysis)

- Mobility and nomadicity

- Accessibility

- Openness

- Transparency (the end-user/application is only concerned with the end-to-end service, in the current Internet this service is the connectivity)

- Neutrality

The architectural properties of the Future Internet shall include:

- Distributed, automated, and autonomy (organic deployment)

- Scalability (e.g., routing scalability $\rightarrow log(n)$ where n is the number of nodes and computational scalability, i.e., to allow support of any business size)

- Resiliency and survivability

- Robustness/stability

- Genericity (e.g., support multiple traffic streams, messages, etc., independent of infrastructure partitioning/divisions, device/system independent)

- Flexibility (e.g., support multiple socioeconomic models, and operational models)

- Simplicity

- Evolvability: evolutionability and extendability

- Heterogeneity (e.g., wireline and wireless access technologies)

- Carbon neutrality

Besides the functional and architectural aspects by themselves, there are also several requirements that the Future Internet should comply with:

- Support for dynamic federation and collaboration for service offerings; this raises the need for publishing and identifying partners to come up with the service offering. This also places requirements related to monitoring and accounting for such dynamic role configurations to effectively work and be profitable.

- Encapsulating off-line and on-line operations; as a result of the vast types of devices as well as user categories, one must support that some devices/services/etc., can be temporarily off-line, e.g., to save power or that links are down.

- Managing risk aspects and evolution incentives; to ensure future evolution, a stepwise approach should be allowed (note: this may come from introducing the virtualization as different "slices" could evolve partly independently).

Based on the current situation and future trend of Internet, researchers could have the following preliminary conclusions [HR08]:

- The future Internet will have an all optical core, consisting of a few hundred optical switches, which provide end-to-end optical paths.

- The role of IP is diminishing; IP will become only an access technology.

- Future Internet = Content + Services + Management [SFD+09]. Since the users will perceive the future Internet in terms of contents and services, the user should no longer be bothered with details such as IP addresses, firewalls etc.

- Access to a mass of sensor data surrounding the users lead to new services that we cannot image today.

- Security and privacy management will become increasingly important.

- Automate management to get the humans as far as possible out of the loop.

- The focus moves from network management, via service management to information and content management.

- The core routing infrastructure of the Internet will be replaced by an all optical switched network (which can be considered as clean-slate design), the focus of research on the future Internet should be on services and the content (which will not need a clean-slate design).

5.10.3   Management Challenges of Future Internet

As a result of the Internet growth and the increasing communication requirements, many patch solutions have been progressively developed and deployed
to enable the Internet to cope with the increasing demand in terms of user
connectivity and capacity. There is, however, a growing consensus among the
scientific and technical community that the current methodology of "patch-
ing" the Internet technology will not be able to sustain its continuing growth
and cope with it at an acceptable cost and speed. Indeed, with the erosion
of the five base design principles (modularization by layering, connectionless
packet forwarding, end-to-end principle, uniform internetworking principle
and simplicity principle), the Internet has progressively become an infrastructure more complex to operate. This complexity results from various layer
violations (e.g., complex cross-layer design) to supposedly optimize network
and system resource consumption, the proliferation of various sublayers, e.g.,
Multi-Protocol Label Switching (MPLS), and Transport Layer Security (TLS)
to expectedly compensate for intrinsic shortcoming in terms of forwarding
performance and security functionality, IP addressing space overload (including network graph locator, node identity, connection termination), and routing system scalability and quality limitations (e.g., Border Gateway Protocol
path exploration and oscillations) to name a few. This complexity progressively impacts the Internet robustness and reliability and in turn impacts its
scalability.

Hence, although the design principles of the Internet are still suitable and
applicable, there is growing evidence that the resulting design components,
as defined today, face certain technical limits (in particular, in terms of scalability). On the other hand, certain objectives of the Internet are no longer
adapted to users' new expectations and behaviors when using the Internet (in
particular, in terms of reliability).

The current Internet architecture is progressively reaching a saturation
point in meeting increasing users' expectations and behaviors as well as progressively showing inability to efficiently respond to new technological challenges (in terms of security, mobility, availability, and manageability), in inability to support the business models necessary to allow value flow in an
increasingly complex service delivery ecosystem that can involve multiple actors, and socioeconomical challenges. Even worse, misguided attempts to sustain the Internet growth resulted into progressive violation and erosion of the
end-to-end principle. Sacrificing the end-to-end principle has in turn resulted
in decreasing the Internet availability, negatively impacting its robustness and
scalability as well as making its manageability more complex. Over time, the
erosion of the end-to-end principle has also resulted in the proliferation of
peer-to-peer and application-specific overlay networks that are progressively
substituting the end-to-end IP networking layer by an end-to-end applicative
communication layer. Indeed, many new applications provide their own path
selection to ensure proper connectivity and quality, resulting in an ineffective

network-level resources use.

From current view of the future Internet, it meets some important technological challenges and management challenges:

1. Routing and addressing scalability and dynamics

2. Resource (forwarding, processing, and storage) and data/traffic manageability and diagnosability

3. Security, privacy, trust, and accountability

4. Availability, ubiquity, and simplicity

5. Adaptability and evolvability to heterogeneous environments, content, context/situation, and application needs (vehicular, ambient/domestic, industrial, etc.)

6. Operating system, application, and host mobility/nomadicity

7. Energy conservation and economic sustainability

8. Managing conflicting interests and dissimilar utility

9. Searchability, localization, selection, composition, and adaptation

10. Beyond just digital communication: semantic (intelligibility of things and content, language, etc.), haptic, emotion, etc.

To meet the management challenges, the overall capabilities of the Future Internet architecture are depicted in Figure 5.32 [GBA08].

5.10.4   Management of Future Internet

Addressing Management

The IPv6 technology has been designed by the IETF to replace the current version of the Internet Protocol, IPv4. This replacement would concurrently re-establish the global end-to-end communication paradigm restoring the valuable properties of the end-to-end IP architecture. Indeed, these properties have been lost in the IPv4 Internet due to the increasing number of Application Layer Gateways (ALGs), Network Address Translators (NATs), and firewalls as well as caches and proxies deployed at various network places and for various applications.

Future Internet

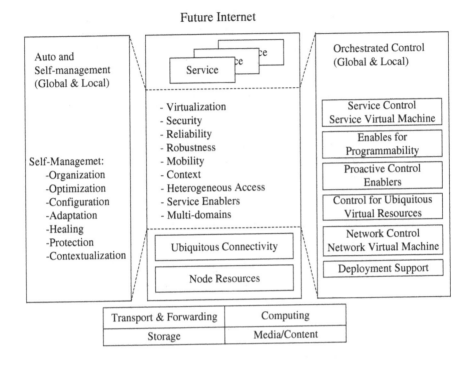

Figure 5.32: Future Internet Capabilities

Security management

- Security

  Security is only supported weakly by the current Internet infrastructure. Internet viruses, phishing, spyware, and identity frauds risk induce reduction of users' confidence in the network and therefore its usefulness. However, the usage of the Internet has partially become the mirror of our "modern" society. As such security is one of the biggest imminent problems facing the Internet.

  For security management, we should consider:

  - Securing the architecture of Future Internet, to have it built-in (security at design time) in addition to the execution (security at running time). This calls for new and innovative approaches such as, for example, collaborative security (leveraging existing ones and research on that field) but also for proper tools to ensure monitoring.

  - Protection against existing and most importantly emerging threats:

1. Means for proactive identification and protection from arbitrary attacks such as Denial of service (DoS) and intrusion detection. DoS attacks are responsible for large changes in traffic characteristics, which may, in turn, significantly reduce the quality of service (QoS) level perceived by all users of the network. This may result in breaking of the service-level agreement, with the Internet Service Provider (ISP) being accountable, potentially causing major financial losses for them.

2. Means for proactive identification and protection from malicious software (malware) such as viruses, spyware, and fraudulent adware.

- Privacy

  Privacy issues fall into two broad categories: users' data privacy and location privacy.

  - Data privacy involves control over personal information contained on the devices and the services providers and in associated database(s).

  - Location privacy involves control over the information regarding the individual's physical location and movement. Major threats caused by location-based services to the user's right of informational self-determination are unsolicited profiling, location tracking, and the disclosure of the user's social network.

It is thus important to define proper global privacy standards on the basis of what to develop the right technology to let people make informed decisions about the services they access. A challenge will also be to preserve anonymity and privacy at large of users of mobile-capable Internet devices, such as mobile Internet, mobile phone, electronic toll payment tags, ePassports, loyalty card programs, mobile RFID (Radio Frequency Identification) service, Mobile P2P Systems, vehicular ad-hoc networks (VANET), Mobile Ad-hoc Networks (MANET).

Identity management is a potential problem because users have to manage multiple identities and credentials, even if they are not actively using all of them. In centralized user identity models, there exists a single identifier and credentials provider that are used by all service providers, either exclusively, or in addition to other identifier and credentials providers. From a user perspective, an increasing number of identifiers and credentials rapidly become totally unmanageable. In the context of FI design, this would call for a user-centric approach to identity management to improve the user experience, and thereby the security of online service provision as a whole. To address these challenges, research in the following space is thus required:

- Design of user interaction for identity management, expressing trustworthiness of identity management to users and privacy-enhancing identity management,

- Accounting/logging tools required for forensic purposes (but not limited to),

- Methodologies and interfaces for managing multiple identities and credentials including delegation,

- Distributed identity management at each providers of services, synchronization with repositories of record, access right framework based on semantic, in particular with respect to user-centric identity and high-level identity assurance.

- Trust

  End-to-end trust is an inclusive approach, where trust is intimately integrated in all the capabilities of Future Internet in a pervasive way to cope with the software for all of the applications. Indeed, in the context of a virtualized environment, the problem is not so much to secure the device by itself but to secure user and/or corporate resources, which are virtualized due to upcoming trends (e.g., resources virtualization, cloud computing).

  To address these challenges, research in the following space is required:

  - Semantics for trust,

  - Trust target certification,

  - Trust life cycle management in highly dynamic environment (modeling, monitoring, audit, recovery),

  - Automated or semi-automated (collaborative) decision-making on trust (including trust negotiation techniques).

Configuration and Monitoring

Configuration management includes the path planning and provisioning and indicates when to establish and release paths. Interdomain path request handling will be a necessary feature and resilience has to ensure that there are different physical paths available. This management technology is well-known, such as TL1, SNMP, GMPLS, and others.

Monitoring is needed for provisioning as well as for security reasons. The main question is what to monitor as Tbps of data will flow through those networks. Monitoring ports are needed for lawful interception / data retention and each country on the path may have different requirements thus monitoring has to be possible at intermediate optical switches.

Heterogeneity Management

The applications built on top of the Future Internet will require a variety of different communication services (one-to-one, one-to-(m)any, many-to-one, and many-to-many): a nonexhaustive list of examples is the following (compared to the so-called best-effort service, most needs can be expressed as a combination of delay and rate):

- real-time service

- guaranteed minimum available bandwidth

- reliable data delivery with relaxed constraints concerning delay and jitter

- data flow resilient against infrastructure failures

- best effort

These different communication services need to be established throughout heterogeneous networks (e.g., wireless, wireline, and hybrid environment) and shall allow an easy integration of future technologies.

Dominant protocols find their roots in the 1980s. However, one observation is the trend not only to use standard IP and Ethernet technology but also to couple those networks tighter with the enterprise. Not only will this create a pressure here for IPv6 migration, the protocols and middleware stacks used in these areas may take advantage from the enhanced IPv6 features such as anycast, mobility, and security. Using these native IP techniques to be more efficient than respective functionality on higher layers when it comes to real-time discovery or security (as IPv6 mandates support of IPSec). Also a seamless integration will simplify network management and monitoring and thus will reduce costs.

The Future Internet has to be an enabler for applications connecting any kind of devices, including also embedded devices, which will be found almost everywhere in future because of computing gets more and more pervasive. Since these embedded devices show special constraints concerning their capabilities, their performance, their energy consumption, the Future Internet concepts, protocols, and services must be built in a way that they are applicable for these ultrasmall embedded systems. E.g., embedded services are required for small microcontrollers to also fulfill the real-time and robustness requirements of some field applications (e.g., in manufacturing, building management, metering infrastructures).

Mobility Management

The FI should "reach" the user also during his/her job or vacancy trips. Up to now, the cost of Internet services offered "on board" is quite high and the quality of the provided service is not particularly exciting. Significant work is thus still to needed to increase the "broadband mobility" for "really mobile" users. The achievement of such ambitious objective can be realized by means of integration between terrestrial and satellite networks in an enhanced "vision of convergence" already mentioned in "4G and beyond" future issues.

– Wireless access: the Internet's main transport protocol (Transmission Control Protocol, or TCP) end-to-end flow control and congestion control needs to cope with corruption and transmission loss and react appropriately (instead of interpreting losses as a sign of congestion). So the key challenge is how to project the needs derived from the existence of heterogeneous links, both wired and wireless yielding a different trade-off between performance, efficiency, and cost.

– TCP connection continuity: using IP address as both network identifier and host identifier but also TCP connection identifier results in TCP connection continuity problem. Resolving the latter requires a certain level of decoupling between the identifier of the position of the mobile host in the network graph (network address) from the identifier used for the TCP connection identification purposes.

- Moving mobile devices such as cellular phones on the Internet is challenging due to limited scalability of Mobile IP (relying on home agent and tunneling). Note that contrary to a persistent belief, the problem is not entirely resolved in IPv6 that still make use of home agents.

- Together with host mobility/nomadicity, suitable localization techniques.

- Take benefit of the radio interface/technologies have inherent broad-/multicast capabilities (air-interface resource consumption).

- Extended broadband coverage to specific critical mobile platforms, such as airplanes, ships, and trains.

Today, mobile networks extensively use roaming agreements to regulate relationships between operators and having subsequent effect on user billing. Future automation of peering/roaming to support a more dynamic infrastructure for both end users and operators shall be further investigated. Aspects like heterogeneous access and even sensor networks accentuate the need.

QoS Management

The incremental and any-purpose usage of Internet resulted in uncontrollable and degraded performance, revealing need for reasonable utilization of network resources. The more heterogeneous and complex the network becomes the more sophisticated solutions it requires to satisfy the QoS. Network operators are facing great challenges of diversified and heterogeneous structure of Future Internet in attempts to provide capable QoS mechanisms and technologies.

The solution can be found in the open and flexible system that would reflect preferences of all participants. Moreover, the system should be automated and responsive to changes by dynamically adjusting the network parameters in case of violations. The main requirements to the architecture of such system would be:

- System heterogeneity and complexity should be transparent to user and all preferences have to be effectively expressed and stored.

- It should be as ubiquitous and modular as possible.

- It should possess easily extensible and pluggable structure.

The main requirements to the functioning of such management system can be formulated as follows:

- Provision of necessary QoS to a particular applications based on the actors policies

- Dynamic adaptation in case of violations based on actors' preferences and/or policies.

Content and Knowledge Management

In FI, content will be much more widely produced. The main challenges are the design of media content by professionals and nonprofessionals supported by open tools for content creation, storage, representation, and indexing ensuring interoperability of various content formats, including efficient search and selection engines, and creation of new innovative media applications.

For distributed media applications, the main challenges are the realization of integrated multicontent communications, integration of classical and new media applications, and creation or adaptation of content dedicated to specific user groups, supported by novel open software and tools for integration of multimedia communications applications.

For new user devices and terminals, the main challenges are associated to advances in integrated, scalable, and modular multimedia devices with auto-configuration and auto-maintenance features and application programming interfaces for new media applications.

Internet of Thing

From the technological point of view, the challenge is to handle the large amount of information coming from the things and to combine it to give useful services. As the current network structure is not suited for this exponential traffic growth, there is a need by all the actors to rethink current networking and storage architectures. It will be imperative to find novel ways and mechanisms to find, to fetch, and to transmit data. Distributed, loosely coupled, ad-hoc peer-to-peer architectures connecting smart devices might represent the network of the future. In this context, the following elements require specific attention:

- Discovery of sensor data in time and space

- Communication of sensor data: Complex Queries (synchronous), Publish/Subscribe (asynchronous)

- Processing of great variety of sensor data streams

- In-network processing of sensor data: correlation, aggregation, filtering

Internet of Services

The term services would include a broad variety of applications that will run over a service-aware made up of elements for which further research is needed:

- Cloud computing: deals with the virtualization of services through more flexible and granular optimization of processing and storage resources, providing applications the necessary run-time support to be provided "as a service" without no limitations of scale in number of users accessing or the amount of resources consumed, all this while complying with the terms of subscribed Service Level Agreements (SLA).

- Open Service Platforms: aim at overcoming incoherent standards, architectures, and deployed service platforms in the Internet. In order to progress toward a coherent "Internet of Services", significant advances need to be made on the interoperability of platforms, their components, core services, APIs, and related open standards. In addition, most of today's Internet service platforms are closed in the sense that they only offer a minimal service interface to the outer world. Open service platforms of the Future Internet will however allow user-designed components and services to be deployed within the platform and therefore lead to intense co-creation involving end users, since they will be able to develop powerful and highly individual services with minimal configuration or programming effort.

- Autonomic computing: aims at creating computer systems capable of self-management to overcome the rapidly growing complexity of computing systems management and to reduce the barrier that that complexity poses to further growth. A general problem of modern distributed computing systems, which has to be considered, is that their complexity, and in particular, the complexity of their management, is becoming a significant limiting factor in their further development. Autonomic computing has to solve the problem of large companies and institutions employing large-scale computer networks for communication and computation. The distributed applications running on these computer networks are diverse and deal with many different tasks, ranging from internal control processes to presenting web content and to customer support.

- Green IT: the need for optimized consumption and efficiency of future platforms is also a significant challenge in the development of new platforms. Indeed, service facilities or data centers concentrate 23% of the overall ICT CO_2 emissions.

Service-Aware Management

Networks are becoming serviceaware. Service awareness means not only that all digital items pertaining to a service are delivered but also that all business or other relations pertaining to a service offer are fulfilled and the network resources are optimally used in the service delivery. In addition, the network's design is moving toward a different level of automation and self-management. The solution is based upon an optimized network and service layers solution that guarantees built-in orchestrated reliability, robustness, mobility, context, access, security, service support, and self-management of the communication resources and services. It suggests a transition from a service agnostic Internet to service-aware network, managing resources by applying Autonomic principles.

In a FIN service management system, a service directory that tells one where to access different types of services, should be capable of determining the availability of service context in a dynamically changing network. New tailor-made services can be provided to consumers by dynamically locating sources of different service context.

As a conclusion, the Internet has for a long time been just good enough, and a clean slate technology alone is not likely to provide sufficient incentives to get better technology deployed. The efficient content distribution and new services taking advantage of a networked world equipped with a large number of sensors will be the main forces driving the deployment of new technologies. Furthermore, great potential in the evolution of the network and service management plane. This is driven by stronger availability and reliability requirements, but also by the fact that users of future Internet services will request new moderation and management services. Network management and new service management functions are becoming a distinguishing and revenue generating factor for many players in the future Internet [SFD+09].

Chapter Review

1. Management challenges and resolution for NGN.

2. Management challenges and resolution for mobile cellular networks and wireless ad-hoc networks.

3. Management challenges and resolution for optical networks.

4. Management challenges and resolution for VPN and P2P networks.

5. Management challenges and resolution for multimedia networks.

6. Management challenges and resolution for satellite networks.

7. Management challenges and resolution for storage networks.

8. Management challenges and resolution for future Internet.

9. Common challenges and resolution for emerging networks.

Chapter 6

Autonomic Computing and Self-Management

Autonomic computing was proposed as a systematic approach to achieving computer-based systems managing themselves without human interventions. An autonomic computing system has four basic characteristics for self-management. They are self-configuration, self-healing, self-optimization, and self-protection. Self-configuration frees people to adjust properties of the system according to changes of the system and environment; self-healing frees people to discover and recover or prevent system failures; self-optimization frees people to achieve best-of-the-breed utilization of resources; and self-protection frees people to secure the system.

6.1 Autonomic Computing

6.1.1 Introduction of Autonomic Computing

Autonomic Computing is an initiative started by IBM in 2001. Its ultimate aim is to develop computer systems capable of self-management, to overcome the rapidly growing complexity of computing systems management, and to reduce the barrier that complexity poses to further growth. In other words, autonomic computing refers to the self-managing characteristics of distributed computing resources, adapting to unpredictable changes while hiding intrinsic complexity to operators and users. An autonomic system makes decisions on its own, using high-level policies; it will constantly check and optimize its status and automatically adapt itself to changing conditions. As widely reported in literature, an autonomic computing framework might be seen composed by Autonomic Components (AC) interacting with each other. An AC can be modeled in terms of two main control loops (local and global) with sensors (for self-monitoring), effectors (for self-adjustment), knowledge and planer/adapter for exploiting policies based on self- and environment awareness.

Driven by such vision, a variety of architectural frameworks based on "self-regulating" autonomic components has been recently proposed. A very similar trend has recently characterized significant research work in the area of multi-agent systems. However, most of these approaches are typically conceived with centralized or cluster-based server architectures in mind and mostly address the need of reducing management costs rather than the need of enabling complex software systems or providing innovative services.

Self-management means different things in different fields: The number of computing devices in use is forecast to grow at 38% per annum and the average complexity of each is increasing. Currently, this volume and complexity is managed by highly skilled humans; but the demand for skilled IT personnel is already outstripping supply, with labor costs exceeding equipment costs by a ratio of up to 18:1. Computing systems have brought great benefits of speed and automation, but there is now an overwhelming economic need to automate their maintenance.

A general problem of modern distributed computing systems is that their complexity, and in particular the complexity of their management, is becoming a significant limiting factor in their further development. Large companies and institutions are employing large-scale computer networks for communication and computation. The distributed applications running on these computer networks are diverse and deal with many different tasks, ranging from internal control processes to presenting web content and to customer support.

Additionally, Mobile computing is pervading these networks at an increasing speed: employees need to communicate with their companies while they are not in their office. They do so by using laptops, PDAs, or mobile phones with diverse forms of wireless technologies to access their companies' data.

This creates an enormous complexity in the overall computer network, which is hard to control manually by one or more human operators. Manual control is time-consuming, expensive, and error-prone. The manual effort needed to control a growing networked computer system tends to increase very quickly. Eighty percent of such problems in infrastructure happen at the client specific application and database layer. Most "autonomic" service providers guarantee only up to the basic plumbing layer (power, hardware, operating system, network, and basic database parameters).

Main application areas in autonomic computing are power management, Grid computing and ubiquitous computing [HM08].

6.1.2 Autonomic Computing Architecture

The goal of an autonomic computing architecture is to limit hands-on intervention to extraordinary situations. Most administrative functions should be carried out according to predefined policies. The autonomic computing architecture is not a technological wonderland but a continuum on which different technologies, organizations, and practitioners find themselves at different times. The road map for the autonomic computing architecture, which is defined by IBM, describes the following five levels of maturity, illustrating how businesses are constantly evolving their IT environment:

Basic → Managed → Predictive → Adaptive → Autonomic

These terms can be defined as:

- Basic: The product and environment expertise resides in human minds, requiring consultation on even mundane procedures.

- Managed: Scripting and logging tools automate routine execution and reporting. Individual specialists review information gathered by the tools to make plans and decisions.

- Predictive: Early warning flags are raised as preset thresholds are tripped. The knowledge base recommends appropriate actions. The proposed resolution of events is leveraged by a centralized storage of common occurrences and experience.

- Adaptive: Building on the predictive capabilities, the adaptive system takes action itself based on the situation.

- Autonomic: Policy drives system activities such as allocation of resources within a prioritization framework.

While examples of predictive and even adaptive systems exist today, the general state of the industry remains at the basic and managed levels. This slows IT reaction times and leads to considerable overhead, duplication of effort, and missed opportunities. At the same time, organizations are looking for extraordinary increases in productivity and contribution from IT. The autonomic computing architecture is poised to solve these problems now and in the future.

6.1.3 Autonomic System

The concept of Autonomic Systems is emerging as a significant new strategic approach to the design of computer-based systems and, generally, refers to computing systems that can manage themselves. Autonomic systems have been recently fostered by the increasing complexity in the design of next-generation, self-organizing, context-aware pervasive computing environments [CCP+06].

The general idea is to create systems that are able to behave autonomously according to high-level description of their objectives. Self-configuration, self-optimization, self-healing, and self-protection have been identified as characterizing the self-management of autonomic systems.

Autonomic systems will be generally constituted by autonomic elements, a myriad of individual entities that dynamically interact with the environment and other elements. Relations among different autonomic elements need to be specified through standard and widely adopted languages and ontologies. Really challenging task will be the understanding of the complex relationships among the local behavior and the global behavior, where the system-wide issues arise from unpredictable relations and from unpredictable scenarios.

A goal-oriented paradigm is adopted, where humans will be in charge only of the high-level specification of autonomic system's goals. The rest of the work will be carried out by the autonomous systems transparently and without human intervention.

A conceptual model of an autonomic system is shown as Figure 6.1.

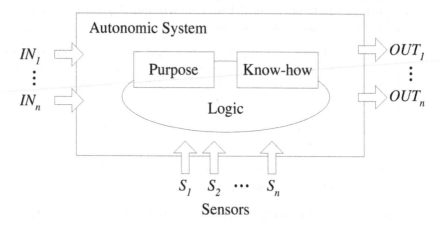

Figure 6.1: Conceptual Model of an Autonomic System

A fundamental building block of an autonomic system is the sensing capability (Sensors S_i), which enables the system to observe its external operational context. Inherent to an autonomic system is the knowledge of the Purpose (intention) and the Know-how to operate itself (e.g., boot-strapping, configuration knowledge, interpretation of sensory data) without external intervention. The actual operation of the autonomic system is dictated by the Logic, which is responsible for making the right decisions to serve its Purpose, and influence by the observation of the operational context (based on the sensor input).

This model highlights the fact that the operation of an autonomic system is purpose-driven. This includes its mission (e.g., the service it is supposed to offer), the policies (e.g., that define the basic behavior), and the "survival instinct." If seen as a control system this would be encoded as a feedback error function or in a heuristically assisted system as an algorithm combined with a set of heuristics bounding its operational space.

Even though the purpose and thus the behavior of autonomic systems vary from system to system, every autonomic system should be able to exhibit a minimum set of properties to achieve its purpose:

- Automatic

  This essentially means being able to self-control its internal functions and operations. As such, an autonomic system must be self-contained and able to start-up and operate without any manual intervention or external help. Again, the knowledge required to bootstrap the system (Know-how) must be inherent to the system.

- Adaptive

  An autonomic system must be able to change its operation (i.e., its configuration, state and functions). This will allow the system to cope with temporal and spatial changes in its operational context either long term (environment customisation/optimisation) or short term (exceptional conditions such as malicious attacks, faults, etc.).

- Aware

  An autonomic system must be able to monitor (sense) its operational context as well as its internal state in order to be able to assess if its current operation serves its purpose. Awareness will control adaptation of its operational behavior in response to context or state changes.

There is a possible solution to enable modern, networked computing systems to manage themselves without direct human intervention. The Autonomic Computing Initiative (ACI) aims at providing the foundation for autonomic systems. It is inspired by the autonomic nervous system of the human body. This nervous system controls important bodily functions (e.g., respiration, heart rate, and blood pressure) without any conscious intervention.

In a self-managing Autonomic System, the human operator does not control the system directly but to define general policies and rules that serve as an input for the self-management process. For this process, IBM has defined the following four functional areas:

- Self-Configuration: Automatic configuration of components;

- Self-Healing: Automatic discovery, and correction of faults;

- Self-Optimization: Automatic monitoring and control of resources to ensure the optimal functioning with respect to the defined requirements;

- Self-Protection: Proactive identification and protection from arbitrary attacks.

An architecture for autonomic systems must accomplish three fundamental goals:

- First, it must describe the external interfaces and behaviors required of individual system components.

- Second, it must describe how to compose these components so that the components can cooperate toward the goals of system-wide self-management.

- Finally, it must describe how to compose systems from these components in such a way that the system as a whole is self-managing.

IBM provides an autonomic computing reference architecture [IBM06]. See Figure 6.2.

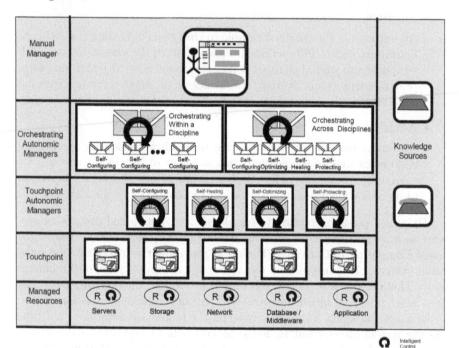

Figure 6.2: Autonomic Computing Reference Architecture

The lowest layer contains the system components or managed resources. These managed resources can be any type of resource (hardware or software) and may have embedded self-managing attributes. The next layer incorporates consistent, standard manageability interfaces for accessing and controlling the managed resources. These standard interfaces are delivered through a manageability end point. Layers three and four automate some portion of the systems process using an autonomic manager.

An autonomic manager is an implementation that automates some management function and externalizes this function according to the behavior defined by management interfaces. Figure 6.3 presents functional details of an autonomic manager [IBM06]. The architecture dissects the loop into four parts that share knowledge. These four parts work together to provide the control loop functionality.

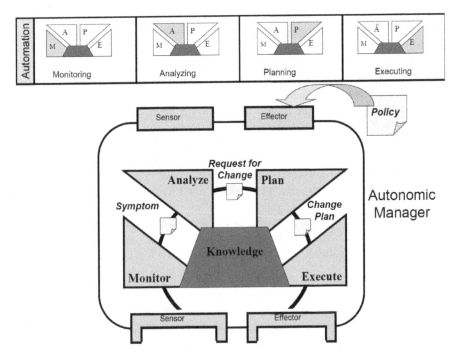

Figure 6.3: Functional Details of an Autonomic Manager

- Monitor function provides the mechanisms that collect, aggregate, and filter, and report details collected from a managed resource.

- Analyze function provides the mechanisms that correlate and model complex situation. These mechanisms allow the autonomic manager to learn about the IT environment and help predict future situations.

- Plan function provides the mechanisms that construct the actions needed to achieve goals and objectives. The planning mechanism uses policy information to guide its work.

- Execute function provides the mechanisms that control the execution of a plan with considerations for dynamic updates.

These four parts communicate and collaborate with one another and exchange appropriate knowledge and data to achieve autonomic management.

6.1.4   Autonomic Networks

The term autonomic is derived from the body's autonomic nerve system, which controls key functions without a conscious awareness or involvement, and the concept of autonomic network is partly borrowed from the world of computing.

The ultimate aim of autonomic networks is to create self-managing networks to overcome the rapidly growing complexity of the Internet and other networks and to enable their further growth, far beyond the size of today.

An autonomic network (computing) should be capable of knowing itself, running itself, adjusting to varying circumstances, and preparing its resources to handle the traffic loads most efficiently. It should be equipped with redundancy in the configurable hardware and with downloadable firmware. When faults happen in the network or when the network is attacked, it should repair the malfunctioning parts and protect itself with minimal or zero human intervention.

The goal of autonomic networks is to increase the automation and reduce human intervention in the management of networks. At the minimum, an autonomic network should have the following three features.

- First, an autonomic network should be self-aware. It should have a detailed knowledge of its elements that includes current states of all the network elements, traffic load across the network, and ultimate capacity, internal network topology, and all connections to other networks. It needs to know the extent of its own resources, including the shared ones among network elements and fixed ones dedicated to certain elements. In a higher level, the network entities in an autonomic network must know both themselves and their surrounding networks, including their activities, and should act accordingly. They should follow and update general rules to interact with neighboring network elements in order to achieve global optimization. They will tap available resources and negotiate the use by other network elements of its underutilized resources, configuring both itself and its connections to other networks in the process.

- Second, an autonomic network should have the capability of self-configuring and self-optimizing. It should never settle for the status quo; it always looks for ways to improve its performance. It should monitor its constituent parts and change global and local network parameters in order to achieve the desired goals of the network operators. An autonomic network must also configure and reconfigure itself under varying conditions. System configuration or setup should occur automatically, as well as dynamic adjustments to that configuration to best handle changing traffic flows and hardware and software resources.

- Third, an autonomic network should be capable of self-healing. It should be able to recover from routine and extraordinary events that might cause some of its parts to malfunction. It should be able to discover faults or potential problems, and then find corrective measures in an automatic manner. These can be activating redundant modules, sending software patches, offloading the processing and traffic load to other network elements, or reconfiguring the system to keep it functioning smoothly. An autonomic network should be an expert in self-protection. It should detect, identify, and protect itself against various types of attacks to maintain overall system security and integrity.

Autognostics is a new paradigm that describes the capacity for computer networks to be self-aware. It is considered as one of the major components of Autonomic Networking. Autognostics includes a range of self-discovery, awareness, and analysis capabilities that provide the autonomic system with a view on high-level state. In metaphor, this represents the perceptual subsystems that gather, analyze, and report on internal and external states and conditions. For example, this might be viewed as the eyes, visual cortex, and perceptual organs of the system. Autognostics, or literally "self-knowledge," provides the autonomic system with a basis for response and validation.

A rich autognostic capability may include many different "perceptual senses." As conditions and states change, they are detected by the sensory monitors and provide the basis for adaptation of related systems. Implicit in such a system are imbedded models of both internal and external environments such that relative value can be assigned to any perceived state. Perceived physical threat can result in rapid shallow breathing related to fight-flight response, a phylogenetically effective model of interaction with recognizable threats.

In the case of autonomic networking, the state of the network may be defined by inputs from:

- individual network elements such as switches and network interfaces including
  - specification and configuration
  - historical records and current state
- traffic flows
- end-hosts
- application performance data
- logical diagrams and design specifications

Most of these sources represent relatively raw and unprocessed views that have limited relevance. Post processing and various forms of analysis must be applied to generate meaningful measurements and assessments against which current state can be derived.

The autognostic system interoperates with:

- configuration management: to control network elements and interfaces

- policy management: to define performance objectives and constraints

- autodefense: to identify attacks and accommodate the impact of defensive responses.

6.2   Context-Aware Management

6.2.1   Context Awareness

Context awareness originated as a term from computer science that sought to deal with linking changes in the environment with computer systems, which are otherwise static. It refers to the idea that computers can both sense and react based on their environment. Devices may have information about the circumstances under which they are able to operate and based on rules, or an intelligent stimulus, react accordingly.

There are three categories of features that a context-aware application can support [Dey01]:

- presentation of information and services to a user

- automatic execution of a service for a user

- tagging of context to information to support later retrieval.

While the computer science community has initially perceived the context as a matter of user location, in the past few years this notion has been considered not simply as a state, but part of a process in which users are involved; thus, sophisticated and general context models have been proposed [BCQ+07], to support context-aware applications which use them to (a) adapt interfaces, (b) tailor the set of application-relevant data, (c) increase the precision of information retrieval, (d) discover services, (e) make the user interaction implicit, or (f) build smart environments.

The way the context is built, managed and exploited [BCQ+07]:

- Context construction: highlights if the context description is built centrally or via a distributed effort; this indicates whether a central, typically design-time, description of the possible contexts is provided, or if a set of partners reaches an agreement about the description of the current context at run-time;

- Context reasoning: indicates whether the context model enables reasoning on context data to infer properties or more abstract context information (e.g., deduce user activity combining sensor readings);

- Context information quality monitoring: indicates whether the system explicitly considers and manages the quality of the retrieved context information, for instance, when the context data are perceived by sensors;

- Ambiguity and incompleteness management: in case the system perceives ambiguous, incoherent, or incomplete context information, indicates if the system can "interpolate" and "mediate" somehow the context information and construct a reasonable "current context";

- Automatic Learning Features: highlights whether the system, by observing the user behavior, individual experiences of past interactions with others, or the environment, can derive knowledge about the context; e.g., by studying the user's browsing habits, the system learns user preferences;

- Multi-Context Modeling: the possibility to represent in a single instance of the model all the possible contexts of the target application, as opposite to a model where each instance represents a context.

6.2.2   Context-Aware Network

A context-aware network is a form of computer network that is a synthesis of the properties of dumb network and intelligent computer network architectures. Dumb networks feature the use of intelligent peripheral devices and a core network that does not control or monitor application creation or operation. Such a network is to follow the end-to-end principle in those applications, which are set up between end peripheral devices with no control being exercised by the network. Such a network assumes that all users and all applications are of equal priority.

In general, context information can be static or dynamic and can come from different network locations, protocol layers and device entities. Any conflict or undesired interaction must be handled by the independent applications. As such the network is most suited to uses in which customization to individual user needs and the addition of new applications are most important. The pure Internet ideal is an example of a dumb network.

An intelligent network, in contrast to a dumb network is most suited for applications in which reliability and stability are of great importance. The network will supply, monitor, and control application creation and operation. A context-aware network is a network that tries to overcome the limitations of the dumb and intelligent network models and to create a synthesis that combines the best of both network models. It is designed to allow for customization and application creation while at the same time ensuring that application operation is compatible not just with the preferences of the individual user but with the expressed preferences of the enterprise or other collectivity which owns the network. The Semantic Web is an example of a context-aware network. Grid networks, pervasive networks, autonomic networks, application-aware networks, service-oriented networks all contain elements of the context-aware model.

In a context-aware network, new applications may be composed from existing network applications. Techniques for modeling applications allow for the identification of applications that satisfy specific functional requirements as well as necessary nonfunctional requirements. This method also allows applications to be described in terms of their overall purposes. For example, an application may describe a business process. The process can be linked to its larger objectives in the organization, including its priority and consequences of failure. The context-aware network can use these descriptions in its function to handle conflict between incompatible applications in the accessing of resources or in the violation of higher-level constraints. The context aware network monitors application operation to ensure that they are compatible with higher-level requirements and constraints and that conflicts are resolved in their light as well.

A context-aware network is suited to applications in which both reliability and the need for system evolution and customization are required. It is finding great purchase in the development of enterprise system for business processes, customer relations management, etc. Service-oriented architectures, which are a specialization of the context-aware model, are the current trend in enterprise computing.

6.3 Self-Management

The essence of autonomic computing systems is self-management, the intent of which is to free system administrators from the details of system operation and maintenance and to provide users with a machine that runs at peak performance 24/7 [KC03] [BBC+03].

Self-management capabilities in a system accomplish their functions by taking an appropriate action based on one or more situations that they sense in the environment. The function of any autonomic capability is a control loop that collects details from the systems and acts accordingly.

Like their biological namesakes, autonomic systems will maintain and adjust their operation in the face of changing components, workloads, demands, and external conditions and in the face of hardware or software failures, both innocent and malicious. The autonomic system might continually monitor its own use, and check for component upgrades, for example.

6.3.1 Self-Configuration

Self-configuration: Automated configuration of components and systems follows high-level policies. Rest of system adjusts automatically and seamlessly. A system is self-configuring to the extent that it automates the installation and setup of its own software in a manner responsive to the needs of the platform, the user, the peer group, and the enterprise. Personal computing often involves user-initiated configuration change, and a self-configuring system understands the implications of these changes and accommodates them automatically.

An autonomic computing system must configure and reconfigure itself under varying (and in the future, even unpredictable) conditions. System configuration or "setup" must occur automatically, as well as dynamic adjustments to that configuration to best handle changing environments. For example, an autonomous computing system must be able to install and set up software automatically. To do so, it will utilize dynamic software configuration techniques, which means applying technical and administrative direction and surveillance to identify and document the functional and physical characteristics of a configurable item. Also to control changes to those characteristics, to record and report change processing and implementation status, and to verify compliance with specified service levels.

The configuration process can be more specifically defined as follows:

- Installation: new installation of necessary components (OS, software, etc.)

- Reconfiguration: reconfiguration of installed components to fit unique situations

- Update: version management of applications or modification of components to correct defects. This also includes re-installation when parts of the configuration files have been corrupted due to virus attack or system error.

A possible way to perform self-configuration in networks is through the application of the PBNM concept, in the sense that network events and conditions can be determinant to trigger automatic reconfiguration of elements, ruled by preconfigured policies.

6.3.2 Self-Healing

Self-healing: System automatically detects, diagnoses, and repairs localized software and hardware problems. A system is self-healing to the extent that it monitors its own platform, detects errors or situations that may later manifest themselves as errors, and automatically initiates remediation. Fault tolerance is one aspect of self-healing behavior, although the cost constraints of personal computing often preclude the redundancy required by many fault-tolerant solutions.

A self-healing network is one that can survive the failure of network entities or links without human intervention. The healing property should be provided at both the physical layer and higher layers. A technique called hot redundancy provides a self-healing property in the physical layer. In this technique, the network nodes are equipped with redundant hardware to receive and process real-time data for key devices or modules. When a device breaks, the redundant one becomes operational instantly without booting and configuring. An IP network provides inherent healing property at the network layer. The failure of a router or leased line in the IP backbone causes the routing protocol to propagate new reachability status throughout the network. All the routers read the information and calculate new routes between end points, which they then use for future packet forwarding. With self-healing capabilities, platforms are able to detect hardware and firmware faults instantly and then contain the effects of the faults within defined boundaries. This allows platforms to recover from the negative effects of such faults with minimal or no impact on the execution of the operating system, middleware, and user-level data [Gu04].

The conventional technology used to perform the basic task of network healing is referred to as fault management and its objective is to detect, isolate, and repair failures in networks. The fault management system proactively diagnoses the cause of abnormal network behavior, and proposes and, if possible, takes corrective actions. Basically, network faults can be classified into hardware and software faults. The effects of such faults vary from underperformance and local traffic congestion to network breakdown.

Examples of hardware faults include the failure of a device due to errors in its logical design, or elements malfunctioning due to simple wear and tear or through external forces such as accidents, acts of nature, mishandling, and vandalism.

Examples of software faults include failure of elements due to incorrect or incomplete design of the software, erratic behavior of elements or the network due to software bugs, and slow or faulty services by the network.

The flow of fault management can be described as follows:

- collect alarms;

- filter and correlate the alarms;

- diagnose faults through analysis and testing;

- determine a plan for correction, display correction options to users, and implement the correction plan;

- verify that the fault is eliminated; and

- record data and determine the effectiveness of the current fault management function.

With the complexity of the heterogeneous ad-hoc networks, more traffic, more nodes, more equipment types, and more protocols will be expected. In order to reduce labor cost and reduce service downtime, maintenance tasks need to be automated. Further, preventive measures should be taken before any serious failure taking place. There are several techniques that have been developed or are being studied in the field of artificial intelligence and, when being integrated properly, they should be able to solve most problems in fault management and help reach the goal of self-healing networks.

6.3.3 Self-Optimization

Self-optimization: Components and systems continually seek opportunities to improve their own performance and efficiency. A system is self-optimizing to the extent that it automatically optimizes its use of its own resources. This optimization must be done with respect to criteria relevant to the needs of a specific user, his or her peer group, and the enterprise. Resource management is one aspect of self-optimizing behavior.

The tasks of the network optimization include both configuration and performance management. The self-optimizing capabilities allow networks to autonomously measure the performance or usage of resources and then tune the configuration of hardware resources to deliver improved performance. Some typical optimization questions that a network manager has include the following: What is the traffic flow in different parts of the network? Is there any congestion taking place? What is the global picture of the quality of services across the network? How can one reduce delay and increase the throughput and the quality of services? There are several measures that are commonly used in the optimization of ad-hoc networks. These include system handover, frequency handover, or relocation, power control, channel switching, and changing antenna parameters such as antenna height and tilt.

The classical network optimization process consists of the following steps:

- information collection: the configuration data and performance measurements are collected.

- data analysis: the collected data are analyzed to determine whether the network is running at its best and if there are any corrective actions to be taken.

- configuration change: instructions are sent to the network elements to make the configuration change.

- verification: a verification test is carried out to ensure that the change of configuration leads to improved performance.

In a self-optimizing network, all these tasks should be accomplished automatically, and the automation makes it easier and faster to respond to the network dynamics.

In a self-optimizing network, new hardware resources are seamlessly integrated with the old ones and configured in a coordinated manner to achieve global optimization. Hardware subsystems and resources can configure and reconfigure autonomously both at boot time and during run time. This action may be initiated by the need to adjust the allocation of resources based on the current optimization criteria or in response to hardware or firmware faults.

Self-optimization also includes the ability to concurrently add or remove hardware resources in response to commands.

6.3.4 Self-Protection

Self-protection: System automatically defends against malicious attacks or cascading failures. It uses early warning to anticipate and prevent system-wide failures. A system is self-protecting to the extent that it automatically configures and tunes itself to achieve security, privacy, function, and data protection goals. This behavior is of very high value to personal computing, which is exposed to insecure networks, an insecure physical environment, frequent hardware, and software configuration changes, and often inadequately trained end users who may be operating under conditions of high stress. Security is one aspect of self-protecting behavior.

Self-protection allows the network to identify, to prevent and to adapt to threats. Firstly, it must integrate security features to network elements including routers, switches, wireless access points, and others network appliances. Secondly, it must allow network elements enabled for security to intercommunicate in a collaborative way, allowing security functions to be extended to equipments of individuals, which eventually connect to other networks bringing risk to the corporate network.

Generally, the main tasks of self-protection:

- Confirm ability to backup and recover data resources

- Network monitoring and IDS (Intrusion Detection System), automatic disconnection of suspicious computers

- Verify that all client machines have latest patches

- Track security advisories.

6.4 Automatic Network Management

Comparing with autonomic, automatic refers to (1)preprogrammed task execution, (2) system works fine until something goes wrong, (3) at latest now, human intervention is needed.

But autonomic refers to (1) self-regulation, (2) system response is also automatic but modulated, (3) system can compensate or work around problems, (4) no human intervention needed.

6.4.1 Network Automation

Network Automation is an intelligent, context-aware software solution that removes the manual human element from network activities, partly or completely automating manual processes, manual configuration activities, and manual policy enforcement. Currently, 45% of network engineers spend their time on manual network activities. By eliminating the need for time-consuming manual actions, Network Automation enables companies to dramatically decrease their IT operating costs while improving the quality of the network, thus resulting in decreased downtime and increased network stability. Network Automation provides organizations with immediate visibility into every detail of their complex, changing IP networks as well as seamless automation of all network maintenance and configuration activities.

Network Automation enables organizations to maximize the value of their IT organization, freeing highly skilled engineers from manual tasks so they can focus on key business initiatives and deliver the quality services the business demands. Functionally, Network Automation delivers a holistic solution, including intelligent change monitoring and management, compliance monitoring and enforcement, security monitoring and network lockdown, vulnerability detection, patch management, inventory management with network discovery and disaster recovery. Network Automation is a fundamental component of all data center management strategies.

Automation can greatly improve the speed and accuracy of complicated, repetitive operations, especially reporting, and better utilize personnel resources to do the work they are best at, such as planning.

For example, automation can help in the following areas:

- detect, prioritize, and fix policy violations

- prevent violations by disallowing activities that contravene defined policies (proactive)

- streamline audit trails

- streamline reporting functions to provide information to decision makers at or near real-time, providing a competitive advantage

- reduce manual effort and thus reduce costs

- reduce errors for repetitive, complex tasks

One important factor of Network Automation is the Centralized Control Model: a system that enforces strict access control and channels all network management activity through tightly controlled processes. In fact, without the Centralized Control Model, a Network Automation system could do drastic damage to the network. The goal of the Centralized Control Model is to provide the required checks and balances for a secure, compliant, stable network without staunching department productivity.

6.4.2 Requirements for Automatic Network Management

An automatic network management solution has the following requirements:

- An Network Automation system has a solid configuration management (CM) system. A robust CM system can be defined as containing the following capabilities:

  - real-time configuration archival and restoration
  - asset management
  - batch script processing
  - compliance validation and enforcement
  - reporting capabilities

- A Network Automation system must provide sophisticated automation capabilities. It is important that the system include rich automation content out of the box, but it is equally important that the system allow users to easily develop their own automation content. The solution must also have mature automation capabilities, including the ability to dynamically manage the results of automation jobs, automate task flow, self-manage required system resources/connections, and automate system actions and behavior. For example, a Network Automation solution can detect that it is running low on disk space or that its FTP server failed to restart properly, and then alert the appropriate party or open a trouble ticket.

- A Network Automation system must encapsulate the Centralized Control Model with capabilities that are easy to use and do not hinder productivity. At the very minimum, it should include:

  - A workflow and approvals engine, capable of modeling complex processes
  - Highly granular permissions model enables device access and user actions
  - Robust permissions management, including notification when user permissions change
  - Centralized access point for all network devices
  - Full keystroke logging for all user/device interaction
  - Network lockdown capabilities
  - Device automation conflict prevention
  - Out-of-the-box integration with other control systems, such as AAA, LDAP/Active Directory or Change Management systems

- A Network Automation system must provide sophisticated redundancy and failover capabilities. Because of the control-oriented nature of the system, redundancy and failover are critical to ensure users do not have to bypass the system.

- A Network Automation system must provide highly flexible and easy to use extensibility features. The system must offer out of the box integration with all major management systems, and an easy process to integrate with home grown systems. These highly flexible extensibility capabilities include the ability to accept dynamic feeds of automation content/system commands, from a website or third-party system.

- The system must deliver an immediate return on investment.

With all of these requirements, the system must still be easy to install and configure, intuitive and straight forward to use. The solution should not lose the efficiencies gained through automation because it, in itself, requires a tremendous deal of maintenance or is cumbersome and difficult to use.

6.4.3 Advantages of Automatic Network Management

Automatic network management is a proactive solution to address the rising network management costs. For the network manager, it delivers the following benefits:

- Increase network uptime and stability

    With real-time change detection, IT dramatically increases visibility into the network situation, precisely who made changes, what they were, and why they happened. In addition, with the automation capabilities, any change can be immediately rolled back to the previous, known good state, decreasing network downtime.

    With the out-of-the-box integration with other network systems, users have better insight into network issues. For example, a configuration change is linked automatically to the specific trouble ticket. Thus, the operator has a complete picture and the issue is faster and easier to resolve.

    Network personnel are able to deploy network-wide configuration changes quickly, reliably, and systematically. They can easily repair configuration errors that are causing a network outage. Reports on network activity provide complete visibility of the IT environment with dynamic, out-of-the-box reports on operational activities, for example, the number of patches deployed in a week or who did what, when, and why.

- Improve and enforce network security

Network managers and security personnel can implement strong user permissions over device access, including by time of day or by specific device. Access privileges for users can be disabled quickly and reliably, without the need to reconfigure every device on the network. Because of the Centralized Control Model, devices can be configured to accept incoming connections from only authorized Network Automation systems, thus decreasing the risk of being hacked by a malicious user. In addition, network manager gain accountability for their team's activity with a keystroke log of actions on each device.

The powerful automation capabilities within the Network Automation solution enable quick response to emerging network threats. For example, the centralized device patch management allows easy deployment and monitoring of device patches, including identifying those OS versions that contain known vulnerabilities.

Real-time enforcement of best-practice configuration standards ensures strict network security standards are obeyed at all times. With Network Automation, users are often automatically prevented from pushing out a configuration change that will violate the defined security standards. In addition, with the advanced network change control workflow and approvals enforcement, mistakes are stopped before they happen, contributing to the overall security and stability of the network. If any security holes do occur, the flexible, powerful notifications immediately alert appropriate parties.

- Compliance validation and enforcement

  Real-time enforcement of best practice configuration standards enforces compliance 24/7. With Network Automation, not just the configuration settings are enforced, but the actual change and workflow processes IT uses to manage the network. Network Automation ensures compliance with policies and best practices by automatically validating proposed changes, deployments, and automatically rolling back unauthorized or noncompliant changes. The granular user permissions ensures that only authorized personnel access devices or are allowed to automate changes on devices.

  Out-of-the-box integration with helpdesk systems provides connection with existing change control processes to ensure that the Network Automation system easily maps into the management ecosystem. As a result, IT personnel do not have to change their workflow making the system easier to adopt.

- Greater automation and lower total cost of ownership

  Network Automation provides the engine to automate multi-step, complex processes. For example, a task to upgrade the software on a device may involve several preliminary steps to determine whether the software upgrade is appropriate for the device. The flexible extensibility with event-triggered actions allows for integration and automation of any job, even jobs that span multiple management systems, for lower total cost of ownership.

The automation engine is capable of executing scripts written in any language to leverage existing automation capabilities already present in the environment.

The built-in device conflict resolution ensures that the automation jobs are accurate and successful. Detailed error reports and analysis allow for swift identification and classification of failures for easy remediation.

Chapter Review

1. What's autonomic computing and autonomic networks?

2. Context-aware management and its advantages.

3. Necessity of self-management and its advantage.

4. Which methods can be used to implement self-configuration, self-healing, self-optimization, and self-protection for networks?

5. What's the difference of autonomic management and automatic management?

Appendix A

Standard Organizations and Sections in Network Management

ANSI	The American National Standards Institute
ARIB	Association of Radio Industries and Business (Japan)
CWTS	China Wireless Telecommunications Standard group (China)
DMTF	Distributed Management Task Force
ETSI	European Telecommunications Standards Institute
IAB	Internet Architecture Board
IEEE	The Institute of Electrical and Electronics Engineers
IETF	The Internet Engineering Task Force
IRTF	Internet Research Task Force
ISO	International Organization for Standardization
ISOC	Internet Society
ITU	International Telecommunication Union
NMRG	The Network Management Research Group
OASIS	Organization for the Advancement of Structured Information Standards
OIF	The Optical Internetworking Forum
OMG	Object Management Group
T1	Standardization Committee-Telecommunications (United States)
TIA	Telecommunications Industry Association (North America)
TMF	Tele Management Forum
TTA	Telecommunications Technology Association (Korea)
TTC	Telecommunications Technology Committee (Japan)
W3C	World Wide Web Consortium

Appendix A

Standard Organizations and Sections in Network Management

Appendix B

SNMPv3 RFCs

RFC 5608: Remote Authentication Dial-In User Service (RADIUS) Usage for Simple Network Management Protocol (SNMP) Transport Models

RFC 5592: Secure Shell Transport Model for the Simple Network Management Protocol (SNMP)

RFC 5591: Transport Security Model for the Simple Network Management Protocol (SNMP)

RFC 5590: Transport Subsystem for the Simple Network Management Protocol (SNMP)

RFC 5345: Simple Network Management Protocol (SNMP) Traffic Measurements and Trace Exchange Formats

RFC 5343: Simple Network Management Protocol (SNMP) Context EngineID Discovery

RFC 4789: Simple Network Management Protocol (SNMP) over IEEE 802 Networks

RFC 4088: Uniform Resource Identifier (URI) Scheme for the Simple Network Management Protocol (SNMP)

RFC 3826: The Advanced Encryption Standard (AES) Cipher Algorithm in the SNMP User-Based Security Model

RFC 3781: Next Generation Structure of Management Information (SMIng) Mappings to the Simple Network Management Protocol (SNMP)

RFC 3584: Coexistence between Version 1, Version 2, and Version 3 of the Internet-Standard Network Management Framework

RFC 3512: Configuring Networks and Devices with Simple Network Management Protocol (SNMP)

RFC 3418: Management Information Base (MIB) for the Simple Network Management Protocol (SNMP)

RFC 3417: Transport Mappings for the Simple Network Management Protocol (SNMP)

RFC 3416: Version 2 of the Protocol Operations for the Simple Network Management Protocol (SNMP)

RFC 3415: View-Based Access Control Model (VACM) for the Simple Network Management Protocol (SNMP)

RFC 3414: User-Based Security Model (USM) for version 3 of the Simple Network Management Protocol (SNMPv3)

RFC 3413: Simple Network Management Protocol (SNMP) Applications

RFC 3412: Message Processing and Dispatching for the Simple Network Management Protocol (SNMP)

RFC 3411: An Architecture for Describing Simple Network Management Protocol (SNMP) Management Frameworks

RFC 3410: Introduction and Applicability Statements for Internet-Standard Management Framework

RFC 2962: An SNMP Application Level Gateway for Payload Address Translation

RFC 2742: Definitions of Managed Objects for Extensible SNMP Agents

Appendix C

ITU-T TMN M.3000 Series for Network Management

M.3000: Tutorial Introduction to TMN

M.3010: Principles for a TMN

M.3020: TMN Interface Specification Methodology

M.3050: Enhanced Telecommunications Operations Map (eTOM)

M.3060: Principles for the Management of the Next-Generation Networks

M.3100: Generic Network Information Model for TMN

M.3200: TMN Management Services Overview

M.3300: TMN Management Capabilities at the F Interface

Appendix D

IEEE 802 Working Group and Executive Committee Study Group

- Active Working Groups and Study Groups

  IEEE 802.1: Higher Layer LAN Protocols Working Group, Link Security Executive Committee Study Group

  IEEE 802.3: Ethernet Working Group

  IEEE 802.11: Wireless LAN Working Group

  IEEE 802.15: Wireless Personal Area Network (WPAN) Working Group

  IEEE 802.15.1: (Bluetooth certification)

  IEEE 802.15.4: (ZigBee certification)

  IEEE 802.16: Broadband Wireless Access Working Group

  IEEE 802.16e: (Mobile) Broadband Wireless Access

  IEEE 802.16.1: Local Multipoint Distribution Service

  IEEE 802.17: Resilient Packet Ring Working Group

  IEEE 802.18: Radio Regulatory TAG

  IEEE 802.19: Coexistence TAG

  IEEE 802.20: Mobile Broadband Wireless Access (MBWA) Working Group

  IEEE 802.21: Media Independent Handoff Working Group

  IEEE 802.22: Wireless Regional Area Networks

- Inactive Working Groups and Study Groups

IEEE 802.2: Logical Link Control Working Group

IEEE 802.5: Token Ring Working Group

- Disbanded Working Groups and Study Groups

IEEE 802.4: Token Bus Working Group

IEEE 802.6: Metropolitan Area Network Working Group

IEEE 802.7: Broadband TAG

IEEE 802.8: Fiber Optic TAG

IEEE 802.9: Integrated Services LAN Working Group

IEEE 802.10: Security Working Group

IEEE 802.12: Demand Priority Working Group

IEEE 802.14: Cable Modem Working Group

IEEE QOS/FC Executive Committee Study Group

- Emerging IEEE 802.11 Standards

IEEE 802.11n: Enhancements for Higher Throughput

IEEE 802.11k: Radio Resource Measurement

IEEE 802.11p: Wireless Access for the Vehicular Environment

IEEE 802.11r: Fast BSS-Transitions

IEEE 802.11s: Wireless Mesh Networks

IEEE 802.11t: Wireless Performance Prediction

IEEE 802.11u: Wireless Inter-working with External Networks

IEEE 802.11v: Wireless Network Management

IEEE 802.11w: Management Frame Protection

IEEE 802.11y: Contention-Based Protocol

Abbreviations

1G	First-Generation Wireless Telephone Technology
2G	Second-Generation Wireless Telephone Technology
3G	Third-Generation Wireless Telephone Technology
3GPP	3rd Generation Partnership Project
4G	Fourth-Generation Wireless Telephone Technology
AA	Active Application
AAA	Authentication, Authorization, and Accounting
ABR	Associative-Based Routing
AC	Autonomic Components
ACI	Autonomic Computing Initiative
ACL	Agent Communication Language
ACO	Ant Colony Optimization
ACSE	Association Control Service Element
ADSL	Asymmetric Digital Subscriber Line
AEE	Agent Execution Environment
AES	Advanced Encryption Standard
AIA	Abstract Intelligent Agents
AIS	Artificial Immune Systems
ALG	Application Layer Gateway
ANN	Artificial Neural Network
ANSI	American National Standards Institute
ANSWER	Automatic Network Surveillance with Expert Rules
AODV	Ad-Hoc On-Demand Distance Vector
API	Application Program Interface
ARIN	American Registry for Internet Numbers
ARP	Address Resolution Protocol
ARPANET	Advanced Research Projects Agency Network
ASN	Abstract Syntax Notation
ATM	Asynchronous Transfer Mode

AuC	Authentication Center
B3G	Beyond 3G
BEEP	Blocks Extensible Exchange Protocol
BEET	Bound End-to-End Tunnel
BER	Bit Error Rate
BGP	Border Gateway Protocol
BML	Business-Management Layer
BN	Bayesian Network
BSP	Basic Support Protocol
BUI	Browser User Interface
C2C	Car-to-Car Communications
CA	Channel Allocation
CAN	Campus Area Network
CBR	Case-Based Reasoning
CDMA	Code Division Multiple Access
CFB	Cipher FeedBack Mode
CGSR	Cluster-Head Gateway Switch Routing
CIDR	Classless Inter-Domain Routing
CIM	Common Information Model
CLI	Command Line Interface
CMIP	Common Management Information Protocol
CMIS	Common Management Information Service
CMISE	Common Management Information Service Element
CMOT	CMIP Over TCP/IP
CoA	Care-of-Address
COPS	Common Open Policy Protocol
CORBA	Common Object Request Broker Architecture
COTS	Commercial Off-the-Shelf
CR	Cognitive Radio
CREN	Corporation for Research and Education Networking
CRN	Cognitive Network
CSP	Constraint Satisfaction Problem
DAG	Directed Acyclic Graph
DAI	Distributed Artificial Intelligence
DAS	Direct Attached Storage
DASD	Direct Access Storage Disk
DCA	Defense Communications Agency
DCC	Data Communications Channel
DCN	Data Communications Network

DDM	Disk Device Module
DDOS	Distributed Denial of Service
DDPS	Dynamic Differentiated Pricing Strategy
DHCP	Dynamic Host Configuration Protocol
DMTF	Distributed Management Task Force
DOC	Distributed Object Computing
DOM	Document Object Model
DOS	Denial Of Service
DPE	Distributed Processing Environment
DSDV	Destination-Sequenced Distance Vector
DSR	Dynamic Source Routing
DS-UWB	Direct Sequence – UWB
DTD	Document Type Definition
DVB	Digital Video Broadcasting
DWDM	Dense Wave Division Multiplexing
EAP	Extensible Authentication Protocol
ECS	Event Correlation Service/Event Correlation System
EDGE	Enhanced Data Rates for GSM Evolution
EE	Execution Environment
EGPRS	Enhanced GPRS
EML	Element Management Layer
ENIAC	Electronic Numerical Integrator and Computer
ESRM	Enterprise Storage Resource Management
eTOM	Enhanced Telecom Operations Map
ETSI	European Telecommunications Standards Institute
FAMA	Fixed Assignment Multiple Access
FCC	Federal Communications Commission
FCIP	Fiber Channel over IP
FDDI	Fiber Distributed Data Interface
FI	Future Internet
FIPA	Foundation for Intelligent Physical Agents
FN	Foreign Network
FSM	Finite-State Machines
FTP	File Transfer Protocol
GAN	Global Area Network
GDH	Group Diffie-Hellman
GDMO	Guidelines for the Definition of Managed Objects
GEO	Geosynchronous Earth Orbit
GFSK	Gaussian Frequency-Shift Keying

GNOC	Global Network Operations Center
GOSIP	US Government OSI Profile
GPRS	General Packet Radio Service
GPS	Global Positioning System
GSM	Global System for Mobile Communications
GTRN	Global Terabit Research Network
GUI	Graphical User Interface
HBA	Host Bus Adapter
HDLC	High-Level Data Link Control
HIP	Host Identity Protocol
HLR	Home Location Register
HMMS	HyperMedia Management Schema
HSDPA	High-Speed Downlink Packet Access
IAB	Internet Architecture Board / Internet Activities Board
ICANN	Internet Corporation for Assigned Names and Numbers
ICMP	Internet Control Message Protocol
IDL	Interface Description Language / Interface Definition Language
IESG	Internet Engineering Steering Group
IETF	Internet Engineering Task Force
IGMP	Internet Group Management Protocol
IMS	IP Multimedia Subsystem
IMT-SC	IMT Single Carrier
InVANET	Intelligent Vehicular Ad-Hoc Network
IP	Internet Protocol
IPFIX	Internet Protocol Flow Information Export
IPsec	Internet Protocol Security
IPTV	Internet Protocol Television
IrCOMM	Infrared Communications Protocol
IrDA	Infrared Data Association
IrFM	Infrared Financial Messaging
IrLAN	Infrared Local Area Network
IrLAP	Infrared Link Access Protocol
IrLMP	Infrared Link Management Protocol
IrOBEX	Infrared Object Exchange
IrPHY	Infrared Physical Layer Specification
IRTF	Internet Research Task Force
ISO	International Organization for Standardization

ISODE	ISO Development Environment
ISP	Internet Service Provider
ITIL	Information Technology Infrastructure Library
ITU	International Telecommunication Union
JPD	Joint Probability Distributions
JVM	Java Virtual Machine
KQML	Knowledge Query Manipulation Language
L2TP	Layer 2 Tunneling Protocol
LAN	Local Area Network
LDAP	Lightweight Directory Access Protocol
LEO	Low Earth Orbit
LNA	Low-Noise Amplifiers
LOS	Line of Sight
LPP	Lightweight Presentation Protocol
LTE	Long Term Evolution
M2M	Machine-to-Machine
MAC	Media Access Control
MAHO	Mobile-Assisted Handoff
MAN	metropolitan area network
MANET	Mobile Ad-Hoc Network
MBR	Model-based Reasoning
MCHO	Mobile-Controlled Handoff
MEO	Medium Earth Orbit
MF	Mediation Function
MIB	Management Information Base
MIP	Mobile IP
MIS	Management Information Systems
MIT	Management Information Tree
MMS	Multimedia Messaging Service
MN	Mobile Node
MNRF	Mobile not reachable flag
MO	Managed Object
MPLS	Multi-Protocol Label Switching
MSC	Mobile-services Switching Center
MSSP	Managed Security Service Provider
n/a	not available / not applicable
NAS	Network Attached Storage
NAT	Network Address Translation
NCHO	Network-Controlled Handoff
NDP	Neighbor Discovery Protocol
NE	Network Elements

NEF	Network Element Functions
NEMO	NEtwork MObility
NETCONF	Network Configuration Protocol
NFC	Near Field Communication
NGN	Next Generation Networking
NMF	Network Management Framework
NML	Network-Management Layer
NMS	Network Management Station/Network Management System
NWG	The Network Working Group
OFDM	Orthogonal Frequency-Division Multiplexing
OMA	Object Management Architecture
OMG	Object Management Group
OO	Object-Oriented
OSC	Optical Supervisory Channel
OSF	Operations System Functions
OSI	Open Systems Interconnection
OSNR	Optical Signal-to-Noise Ratio
OSPF	Open Shortest Path First
OSS	Operations Support System
P2P	Peer-to-Peer
PAN	Personal Area Network
PAR	Project Authorization Request
PBM	Policy-Based Management
PBNM	Policy-Based Network Management
PCS	Personal Communication Services
PDA	Personal Digital Assistant
PDP	Policy Decision Points
PDU	Protocol Data Unit
PEP	Policy Enforcement Points
PFSM	Probabilistic Finite-State Machines
PGP	Pretty Good Privacy
PMTUD	Path Maximum Transmission Unit Discovery
PPTP	Point-to-Point Tunneling Protocol
PSO	Particle Swarm Optimization
PU	Primary User
QAF	Q Adaptor Function
QoS	Quality of Service
RADIUS	Remote Authentication Dial In User Service

RFC	Request for Comments
RFID	Radio Frequency Identification
RIAA	Recording Industry Association of America
RMI	Remote Method Invocation
RMON	Remote MONitoring
ROI	Return on Investment
ROSE	Remote Operation Service Element
RPC	Remote Procedure Call
SAN	Storage Area Networks
SCM	Software Configuration Management
SDH	Synchronous Digital Hierarchy
SDS	Stochastic Diffusion Search
SI	Swarm Intelligence
SLA	Service Level Agreement
SLAAC	Stateless Address Auto Configuration
SLS	Service Level Specification
SM	Situation Management
SMI	Structure of Management Information
SML	Service-Management Layer
SMS	Short Message Service
SNMP	Simple Network Management Protocol
SOA	Service Oriented Architecture
SOAP	Simple Object Access Protocol
SOC	Self-Organized Criticality
SONET	Synchronous Optical Networks
SSL	Secure Sockets Layer
SU	Secondary user
TACACS	Terminal Access Controller Access Control System
TCL	Tool Command Language
TCP	Transmission Control Protocol
TDMA	Time Division Multiple Access
TLS	Transport Layer Security
TMN	Telecommunications Management Network
TMSI	Temporary Mobile Subscriber Identity
TORA	Temporally Ordered Routing Algorithm
TTL	Time to Live
UDP	User Datagram Protocol
UIFN	Universal International Freephone Numbers

UMTS	Universal Mobile Telecommunications System
USM	User-Based Security Model
UWB	Ultra-wide Band
VAN	Virtual Active Network
VANET	Vehicular Ad Hoc Networks
VLR	Visitor Location Register
VoIP	Voice over Internet Protocol
VPN	Virtual Private Network
VTOC	Volume Table of Content
VVDS	VSAM VTOC Data Set
WAN	Wide Area Network
WANET	Wireless Ad-Hoc Network
WAP	Wireless Application Protocol
WBEM	Web-Based Enterprise Management
WebNM	Web-Based Network Management
WEP	wired equivalent privacy
Wi-Fi	Wireless Fidelity
WiMax	Worldwide Interoperability for Microwave Access
WLAN	Wireless Local Area Network
WMAN	Wireless Metropolitan Area Network
WMN	wireless mesh network
WPAN	Wireless Personal Area Network
WRP	Wireless Routing Protocol
WSDL	Web Service Description Language
WSF	Work Station Functions
WSN	Wireless Sensor Network
WWW	World Wide Web
XML	Extensible Markup Language
XPath	XML Path Language
XSL	Extensible Stylesheet Language
XSLT	XSL Transformations

Bibliography

[AA06] M. Gökçen Arslan and Fatih Alagöz. Security Issues and Performance Study of Key Management Techniques over Satellite Links. Proceedings of 11th Intenational Workshop on Computer-Aided Modeling, Analysis and Design of Communication Links and Networks, Pages 122 – 128, 2006.

[AC07] Hüseyin Arslan and Hasari Celebi. Location Information Management Systems for Cognitive Wireless Networks. H. Arslan (Ed.), Cognitive Radio, Software Defined Radio, and Adaptive Wireless Systems, Pages 291 – 323. Springer, 2007.

[AD99] E. Aboelela and C. Douligeris. Fuzzy Temporal Reasoning Model for Event Correlation in Network Management. 24th Conference on Local Computer Networks, LCN'99, Pages 150 – 159, 1999.

[ALV+06] Ian F. Akyildiz, Won-Yeol Lee, Mehmet C. Vuran, and Shantidev Mohanty. NeXt generation/dynamic spectrum access/cognitive radio wireless networks: A survey. Computer Networks 50 (2006) 2127 – 2159, 2006.

[ALV+08] Ian F. Akyildiz, Won-Yeol Lee, Mehmet C. Vuran, and Shantidev Mohanty. A Survey on Spectrum Management in Cognitive Radio Networks. IEEE Communications Magazine, April 2008, Pages 40 – 48.

[AMH+99] I. F. Akyildiz, J. McNair, J. S. M. Ho, H. Uzunalioglu, and Wenye Wang. Mobility management in next-generation wireless systems. Proceedings of the IEEE, Volume 87, Issue 8, Aug. 1999 Pages 1347 – 1384.

[ANS94] ANSI T1.215 OAM&P – Fault Management Messages for Interface between Operations Systems and Network Elements, 1994.

[AP93] S. Aidarous and T. Plevyak. Telecommunications Network Management into the 21st Century, IEEE Press, Piscataway, NJ, 1993.

[ATIS01] ATIS Telecom Glossary 2000, ATIS Committee T1A1 Performance and Signal Processing (approved by the American National Standards Institute), 28 February 2001.

[BA07] Raouf Boutaba and Issam Aib. Policy-based Managment: A Historical Perspective. Journal of Network and System Management. 15: 447 – 480, 2007.

[BBC+03] D. F. Bantz, C. Bisdikian, D. Challener, J. P. Karidis, S. Mastrianni, A. Mohindra, D. G. Shea, and M. Vanover. Autonomic personal computing. IBM Syst. J. 42, 1, 165 – 176, 2003.

[BBD+07] Sasitharan Balasubramaniam, Dmitri Botvich, William Donnelly, Mícheál ó Foghlú, and John Strassner. Bio-inspired Framework for Autonomic Communication Systems. Studies in Computational Intelligence (SCI) 69, 3 – 19, Springer, 2007.

[BBJ+09] Sasitharan Balasubramaniam, Dmitri Botvich, Brendan Jennings, Steven Davy, William Donnelly, and John Strassner. Policy-constrained bio-inspired processes for autonomic route management. Computer Networks, Volume 53, Issue 10, Pages 1666 – 1682, 2009.

[BBM+93] S. Brugnoni, G. Bruno, R. Manione, E. Montariolo, E. Paschetta, and L. Sisto. An expert system for real time fault diagnosis of the Italian telecommunications network. In Proc. IFIP/IEEE International Symposium on Integrated Network Management, Pages 617 – 628, Apr. 1993.

[BCF94] A. Boulouts, S. Calo, and A. Finkel. Alarm Correlation and Fault Identification in Communication Networks. IEEE Transactions on Communication, Vol. 42(2 – 4), Pages 523 – 533, 1994.

[BCF+95] A. T. Bouloutas, S. B. Calo, A. Finkel, and I. Katzela. Distributed Fault Identification in Telecommunication Networks. Journal of Network and Systems Management, 3(3):295 – 312, 1995.

[BCF+02] A. Barone, P. Chirco, G. Di Fatta, and G. Lo Re. A management architecture for active networks. Proceedings of the 4th Annual International Workshop on Active Middleware Services, Pages 41 – 48, 2002.

[BCQ+07] Cristiana Bolchini, Carlo A. Curino, Elisa Quintarelli, Fabio A. Schreiber, and Letizia Tanca. A data-oriented survey of context models. SIGMOD Rec. (ACM) 36 (4): 19 – 26, 2007.

[BDV97] Kenneth Basye, Thomas Dean and Jeffrey Scott Vitter. Coping with uncertainty in map learning. Machine Learning 29(1): 65 – 88, 1997.

[BH95] Lisa Burnell and Eric Horvitz. Structure and chance: Melding logic and probability for software debugging. Communications of the ACM, 38(3): 31 – 41, 57, Mar. 1995.

[BHS93] A. Bouloutas, G. Hart, and M. Schwartz. Fault Identification Using a Finite State Machine Model with Unreliable Partially Observed Data Sequences. IEEE Transactions on Communications. Vol. 41, No. 7, Pages 1047 – 1083, July 1993.

[BK02] Stephen F. Bush and Amit B. Kulkarni. Active Network and Active Network Management: A Proactive Management Framework. ISBM: 0-306-46560-4, Kluwer Academic Publishers, 2002.

[BKR99] J. Bredin, D. Kotz, and D. Rus. Economic Markets as a means of open Mobile-Agent Systems. Proceedings of the Workshop Mobile Agents in the Context of Competition and Cooperation (MAC3) at Autonomous Agents '99, Pages 43 – 49, May 1999. ACM press.

[BMI+00] J. Bredin, R. T. Maheswaran, C. Imer, T. Basar, D. Kotz, and D. Rus. A Game-Theoretic Formulation of Multi-Agent Resource Allocation. Proceedings of the 4th International Conference on Autonomous Agents, Pages 349 – 356, 2000.

[BP02] R. Boutaba and A. Polyrakis. Projecting Advanced Enterprise Network and Service Management to Active Networks. IEEE Network, 16(1): Pages 28 – 33, 2002.

[Bru00] M. Brunner. Active Networks and its Management. Proceedings of 1st European Conference on Universal Multiservice Networks, Pages 414 – 424, 2000.

[BSF07] Remi Badonnel, Radu State, Olivier Festor. Management of Ad-Hoc Networks. Handbook of Network and System Administration, Jan Bergstra and Mark Burgess (Eds.), ISBN: 978-0-444-52198-9, Elsevier B.V., 2007.

[Bun96] W. Buntine. Graphical Models for Discovering Knowledge. P. Smyth and R. Uthurusamy (Eds.), Advances in Knowledge Discovery and Data Mining, Pages 59 – 82, 1996. AAAI/MIT Press.

[BX02] Raouf Boutaba and Jin Xiao. Network Management: State of the Art. IFIP World Computer Congress, Pages 127 – 146, 2002.

[CCP+06] Iacopo Carreras, Imrich Chlamtac, Francesco De Pellegrini, Csaba Kiraly, Daniele Miorandi, and Hagen Woesner. A Biological Approach to Autonomic Communication Systems. Trans. on Comput. Syst. Biol. IV, LNBI 3939, Pages 76 – 82, 2006.

[CDL+99] R. G. Cowell, A. P. Dawid, S. L. Lauritzen, and D. J. Spiegelhalter. Probabilistic Networks and Expert Systems. New York: Springer-Verlag, 1999.

[CG06] H.-H. Chen and M. Guizani. Next generation wireless systems and networks. John Wiley & Sons, ISBN-13: 978-0-470-02434-8, 2006.

[CH96] Jiann-Liang Chen and Pei-Hwa Huang. A fuzzy expert system for network fault management. IEEE International Conference on Systems, Man, and Cybernetics, 1996, Volume 1, Pages 328 – 331, 1996.

[CHJ03] Mi-Jung Choi, James W. Hong, and Hong-Taek Ju. XML-Based Network Management for IP Networks. ETRI Journal, Volume 25, Number 6, Pages 445 – 463, 2003.

[Cle06] Alexander Clemm: Network Management Fundamentals. CiscoPress, ISBN: 1-58720-137-2, 2006.

[CMP89] A. A. Covo, T. M. Moruzzi, and E. D. Peterson. AI-assisted telecommunications network management. Proceedings of IEEE Global Telecommunications Conference (GLOBECOM89), Pages 487 – 491, 1989.

[Coo90] G. Cooper. Computational complexity of probabilistic inference using Bayesian belief networks. Artificial Intelligence, 42: 393 – 405, 1990.

[CPP+08] K.-C. Chen, Y.-J. Peng, N. Prasad, Y.-C. Liang, and S. Sun. Cognitive radio network architecture: part II – trusted network layer structure. Proceedings of the 2nd International Conference on Ubiquitous Information Management and Communication, Pages 120 – 124, 2008.

[CS03] Seraphin Calo and Morris Sloman. Policy-Based Management of Networks and Services. Journal of Network and Systems Management, Vol. 11, No. 3, September 2003.

[CSK01] R. Chen, K. Sivakumar, and H. Kargupta. Distributed Web Mining Using Bayesian Networks from Multiple Data Streams, Proceedings IEEE International Conference on Data Mining, Pages 75 – 82, 2001.

[CZL+04] Mike Chen, Alice X. Zheng, Jim Lloyd, Michael I. Jordan, and Eric Brewer. Failure Diagnosis Using Decision Trees. Proceedings of the First International Conference on Autonomic Computing, Pages 36 – 43, 2004.

[Dag01] T. Dagiuklas. NGN Architecture and Characteristics. ETSI, June 2001.

[Dav08] Klaus David (Ed.). Technologies for the Wireless Future, Volume 3, ISBN: 9780470993873, Wireless World Research Forum (WWRF), John Wiley & Sons, 2008.

[Dey01] Anind K. Dey. Understanding and Using Context. Personal Ubiquitous Computing 5(1): 4 – 7, 2001.

[Dha07] S. Dhar, Applications and Future Trends in Mobile Ad Hoc Networks, Business Data Communications and Networking: A Research Perspective, Volume 1, edited by Jairo Gutiérrez, Idea Group, Pages 272 – 300, 2007.

[Din07] Jianguo Ding. Probabilistic Management of Distributed Systems. Bernd J. Krämer and Wolfgang A. Halang (Eds.), Contributions to Ubiquitous Computing, Pages 235 – 262, ISBN: 3-540-44909-4, Springer, 2007.

[Din08] Jianguo Ding. Probabilistic Fault Management in Distributed Systems. ISSN: 0178-9627, ISBN: 978-3-18-379110-1, VDI Verlag, Germany, 2008.

[DKB+05] Jianguo Ding, Bernd Krämer, Yingcai Bai, and Hansheng Chen. Backward Inference in Bayesian Networks for Distributed Systems Management. Journal of Network and Systems Management, 13(4): 409 – 427, Dec 2005.

[DL07] T. G. Dietterich and P. Langley. Machine learning for cognitive networks: Technology assessment and research challenges. In Q. Mahmoud (Ed.), Cognitive Networks: Towards Self-Aware Networks. New York: John Wiley, 2007.

[DLC03] Timon C. Du, Eldon Y. Li, and An-Pin Chang. Mobile Agent in Distributed Network Management. Communications of the ACM, Volume 46, Issue 7, Pages 127 – 132, 2003.

[DLW93] R. H. Deng, A. A. Lazar and W. Wang. A probabilistic Approach to Fault Diagnosis in Linear Lightwave Networks, IEEE Journal on Selected Areas in Communications, Vol. 11, no. 9, Pages 1438 – 1448, December 1993.

[DM89] J. E. Dobson and J. A. McDermid. A Framework for Expressing Models of Security Policy. IEEE Symposium on Security & Privacy, Pages 229 – 241, 1989.

[DMP+08] Franco Davoli, Norbert Meyer, Roberto Pugliese, and Sandro Zappatore. Grid Enabled Remote Instrumentation, ISBN:978-0-387-09662-9, Springer Science+Business Media, 2008.

[DSW+91] A. Dupuy, S. Sengupta, O. Wolfson, and Y. Yemini. Design of the Netmate Network Management System. I. Krishnan, W. Zimmer (Eds.), Integrated Network Management II, North-Holland, Amsterdam, Pages 639 – 650, 1991.

[DV95] Gabi Dreo and Robert Valta. Using Master Tickets as Storage of Problem-Solving Expertise. Proceedings of IFIP/IEEE International Symposium on Integrated Network Management, IV, Pages 328 – 340, 1995.

[FJM+98] G. Forman, M. Jain, J. Martinka, M. Mansouri-Samani, and A. Snoeren. Automated End-to-end System Diagnosis of Networked Printing Services Using Model Based Reasoning. A.S. Sethi (Ed.), 9th International Workshop on Distributed Systems: Operations and Management, Pages 142 – 154, 1998.

[FKW96] A. S. Franceschi, L. F. Kormann, and C. B. Westphall. Performance Evaluation for Proactive Network Management. Proceedings of the ICC'96 International Conference on Communications, vol. I, Pages 22 – 26, 1996.

[GBA08] Alex Galis, Marcus Brunner, Henrik Abramowicz. Position Paper: Management and Service-aware Networking Architectures (MANA) for Future Internets, 2008.

[GF01] Hugo Gamboa and Ana Fred. Designing Intelligent Tutoring Systems: A Bayesian Approach. Proceedings of 3rd International Conference on Enterprise Information Systems, Pages 452 – 458, 2001.

[GH96] R. D. Gardner and D. A. Harle. Methods and Systems for Alarm Correlation. Proceedings of GLOBECOM96, Pages 136 – 140, 1996.

[GH98] R. D. Gardner and D. A. Harle. Pattern Discovery and Specification Techniques for Alarm Correlation. Proceedings of Network Operation and Management Symposium (NOMS '98), Pages 713 – 722, 1998.

[Gia07] Giovanni Giambene (Ed.). Resource Management in Satellite Networks Optimization and Cross-Layer Design. ISBN: 978-0-387-36897-9, Springer, 2007.

[GKO+96] D. W. Guerer, I. Khan, R. Ogler, and R. Keffer. An Artificial Intelligence Approach to Network Fault Management, SRI International, 1996.

[GS05] David Groth and Toby Skandier. Network+ Study Guide, Fourth Edition'. Sybex, Inc.. ISBN 0-7821-4406-3, 2005.

[GSP00] Jairo A. Gutiérrez, Donald P. Sheridan, and R. Radhakrishna Pillai. A Framework and Lightweight Protocol for Multimedia Network Management. Journal of Network and Systems Management,Volume 8, Number 1 / March, Pages 33 – 48, 2000.

[Gu04] Y. Jay Guo. Advances in mobile radio access networks. ISBN 1-58053-727-8, Artech House, 2004.

[Gu06] Ankur Gupta. Network Management: Current Trends and Future Perspectives, Journal of Network and Systems Management, Volume 14, Number 4 / December, Pages 483 – 491, 2006.

[HAN99] H. Hegering, S. Abeck, and B. Neumair. Integrated Management of Network Systems. Morgan Kaufmann Publishers, 1999.

[HB05] Markus Hofmann and Leland R. Beaumont. Content networking: architecture, protocols, and practice. Morgan Kaufmann, ISBN: 1558608346, 2005.

[HBR95] D. Heckerman, J. Breese, and K. Rommelse. Decision-theoretic Troubleshooting. Communications of the ACM, 38:49 – 57, 1995.

[HC02] Chih-Lin Hu and Wen-Shyen E. Chen. A Mobile Agent-Based Active Network Architecture for Intelligent Network Control. Journal of Information Sciences, Volume 141, Pages 3 – 35, 2002.

[HCF95] K. Houck, S. Calo, and A. Finkel. Towards a practical alarm correlation system. A. S. Sethi, F. Faure-Vincent, and Y. Raynaud (Eds.), Integrated Network Management IV, Chapman and Hall, London, Pages 226 – 237, 1995.

[Hei27] W. Heisenberg. Ueber den anschaulichen Inhalt der quantentheoretischen Kinematik und Mechanik. Zeitschrift für Physik 43 Pages 172 – 198, 1927. English translation in (Wheeler and Zurek, 1983), Pages 62 – 84.

[Hek06] R. Hekmat. Ad-Hoc Networks: Fundamental Properties and Network Topologies. Dordrecht, Springer, 2006.

[HJ97a] C. Hood and C. Ji. Automated Proactive Anomaly Detection. Proceedings of IEEE International Conference of Network Management (IM97), Pages 688 – 699, May 1997.

[HL06] James W. Haworth and John A. Liffrig. Remote Network Management for Optical Networks. Bell Labs Technical Journal 11(2), 191 – 201, 2006.

[HM08] Markus C. Huebscher and Julie A. McCann. A Survey of Autonomic Computing – Degrees, Models, and Applications. ACM Computing Surveys, Vol. 40, No. 3, Article 7, 2008.

[How03] How Much Information? 2003. http://www.sims.berkeley.edu/research/projects/how-much-info/.

[HR08] Iris Hochstatter and Gabi Dreo Rodosek. Management Challenges for Different Future Internet Approaches. Proceedings of the Poster and Demonstration Paper Track of the 1st Future Internet Symposium (FIS'08), Pages 17 – 19, 2008.

[HSV99] M. Hasan, B. Sugla, and R. Viswanathan. A Conceptual Framework for Network Management Event Correlation and Filtering Systems. M. Sloman, S. Mazumdar, and E. Lupu (Eds.), IEEE Integrated Network Management VI, Pages 233 – 246, 1999.

[IBM06] IBM. An Architecture Blueprint for Autonomic Computing. White paper. 4th Edition, 2006.

[ISC] https://www.isc.org/.

[ISO93] ISO DIS 10165-1: Information Processing Systems - Open Systems Interconnection - Structure of Management Information - Part 1: Management Information Model, Geneva, 1993.

[ITU92] ITU-T. Recommendation X. 700: Management Framework for Open Systems Interconnection (OSI) for CCITT applications, September 1992.

[ITU96] ITU-T. Recommendation M.3010: Principles for a Telecommunications Management Network, May 1996.

[ITU2012] ITU-T Rec. Y.2012.

[IWS] http://www.internetworldstats.com/.

[JBL07] Gabriel Jakobson, John Buford, and Lundy Lewis. Situation Management: Basic Concepts and Approaches. Information Fusion and Geographic Information Systems. Pages 18 – 33, Springer, 2007.

[Jep03] Thomas C. Jepsen. Distributed Storage Networks Architecture, Protocols and Management. ISBN 0-470-85020-5, John Wiley & Sons, 2003.

[JL04] Xiaolong Jin and Jiming Liu. From Individual Based Modeling to Autonomy Oriented Computation. Matthias Nickles, Michael Rovatsos, and Gerhard Weiss (Eds.), Agents and Computational Autonomy: Potential, Risks, and Solutions, Pages 151 – 169, Lecture Notes in Computer Science, vol. 2969, 2004.

[JLM+05] G. Jakobson, L. Lewis, C. J. Matheus, M. M. Kokar,J. Buford. Overview of situation management at SIMA 2005. Proceedings of IEEE Military Communications Conference, 2005 (MILCOM 2005), Pages 1630 – 1636.

[JP93] J. F. Jordaan and M. E. Paterok. Event correlation in Heterogeneous Networks Using the OSI Management Framework. H. G. Hegering and Y. Yemini (Eds.), Integrated Network Management III, North-Holland, Amsterdam, Pages 683 – 695, 1993.

[JW93] G. Jakobson and M. Weissman. Alarm Correlation. IEEE Network. Vol. 7, No. 6, Pages 52 – 59, Nov. 1993.

[JW95] G. Jakobson and M. D. Weissman. Real-time Telecommunication Network Management: Extending Event Correlation with Temporal Constraints. A. S. Sethi, F. Faure-Vincent, and Y. Raynaud (Eds.), Integrated Network Management IV, Chapman and Hall, London, Pages 290 – 302, 1995.

[Kat96] S. Kätker. A modeling framework for integrated distributed systems fault management. C. Popien (Ed.), Proceedings of IFIP/IEEE Internat. Conference on Distributed Platforms, Pages 187 – 198, 1996.

[KC03] Jeffrey O. Kephart, David M. Chess. The Vision of Autonomic Computing. Computer, Volume: 36, Issue: 1, Pages 41 – 50, 2003.

[KF07] Marcos D. Katz and Frank H. P. Fitzek. Cooperative and Cognitive Networks: A Motivating Introduction. F. H. P. Fitzek and M. D. Katz (Eds.), Cognitive Wireless Networks. ISBN: 978-1-4020-5978-0 Springer, 2007.

[KG97] S. Kätker and K. Geihs. A Generic Model for Fault Isolation in Integrated Management System. Journal of Network and Systems Management, Special Issue on Fault Management in Communication Networks, Vol. 5, No. 2, June 1997.

[KK94] Harald Kirsch and Khristian Kroschel. Applying Bayesian Networks to Fault Diagnosis. Proceedings of 3rd IEEE conference on Control Applications, Pages 895 – 900, 1994.

[KK98] Bernd J. Krämer, Thomas Koch. Distributed Systems Management Software-in-the-Loop. International Journal of Software Engineering and Knowledge Engineering, Vol. 8, No. 1, Pages 55 – 76, 1998.

[KKR95] T. Koch, B. Kramer, and G. Rohde. On a Rule Based Management Architecture. Proceedings of the 2nd International Workshop on Services in Distributed and Networked Environment, Pages 68 – 75, June 1995.

[Kle08] L.Kleinrock. History of the internet and its flexible future. IEEE Wireless Communications, Volume: 15, Issue: 1, Pages 8 – 18, 2008.

[KM08] M. Kajko-Mattsson and C. Makridis. Outline of an SLA Management Model. Proceedings of 12th European Conference on Software Maintenance and Reengineering, Pages 308 – 310, 2008.

[KMT99] M. Klemettinen, H. Mannila, and H. Toivonen. Rule Discovery in Telecommunication Alarm Data, Journal of Network and Systems Management, 7(4), Pages 395 – 423, 1999.

[Koh78] Z. Kohavi. Switching and Finite Automata Theory, New York, NY: McGraw-Hill, 1978.

[Kol02] Michael O. Kolawole. Satellite Communication Engineering. ISBN: 0-8247-0777-X, Marcel Dekker, 2002.

[Koz05] Charles M. Kozierok. The TCP/IP Guide: A Comprehensive, Illustrated Internet Protocols Reference, No Starch Press, ISBN 9781593270476, 2005.

[KP97] S. Kätker and M. Paterok. Fault Isolation and Event Correlation for Integrated Fault Management. A. Lazar, R. Sarauo, and R. Stadler (Eds.), Integrated Network Management V, Chapman and Hall, London, Pages 583 – 596, 1997.

[KS95] I. Katzela and M. Schwarz. Schemes for Fault Identification in Communication Networks. IEEE/ACM Transactions on Networking, vol. 3, Pages 753 – 764, 1995.

[KTT+05] Curtis M. Keliiaa, Jeffrey L. Taylor, Lawrence F. Tolendino, et al. Policy Based Network Management: State of the Industry and Desired Functionality for the Enterprise Network, Security Policy / Testing Technology Evaluation. Tech Report: SAND2004-3254, 2005.

[KV97] G. Kumar and P. Venkataram, Artificial intelligence approaches to network management: recent advances and a survey. Computer Communications, Vol. 20 No. 1, Pages 1313 – 1322, 1997.

[KW+04] Fernando Koch, Carlos Becker Westphall, et al. Distributed Artificial Intelligence for Network Management Systems – New Approaches, LNCS 3126, Pages 135 – 145, 2004.

[KYY+95] S. Kliger, S. Yemini, Y. Yemini, D. Oshie, and S. Stolfo. A Coding Approach to Event Correlation. Proceedings of the Fourth IEEE/IFIP International Symposium on Integrated Network Management, Pages 266 – 277, Chapman and Hall, London, UK, May 1995.

[KZ96] W. Klosgen and J. M. Zytkow. Knowledge Discovery in Database Terminology. U. M. Fayyad, G. Piatetsky-Shapiro, P. Smyth, and R. Uthurusamy (Eds.), Advances in Knowledge Discovery and Data Mining, Pages 573 – 592. AAAI Press/The MIT Press, 1996.

[LCW+97] W. Liu, C. Chiang, H. Wu, C. Gerla. Routing in Clustered Multihop Mobile Wireless Networks with Fading Channel. In Proc. IEEE SICON'97, Pages 197 – 211, 1997.

[LD93] L. Lewis and G. Dreo. Extending Trouble Ticket Systems to Fault Diagnosis. IEEE network, Pages 44 – 51, Nov. 1993.

[Le05] F. L. Lewis. Wireless Sensor Networks. Smart Environments: Technologies, Protocols, Applications, Chapter 2, ed. D. J. Cook and S. K. Das, Wiley, New York, ISBN: 0471544485, 2005.

[Leu02] K. Leung. Diagnostic Cognitive Assessment and E-learning. Proceedings of World Conference on E-Learning in Corporate, Government, Healthcare, and Higher Education, Pages 578 – 592, 2002.

[Lew93] L. Lewis. A case-based reasoning approach to the resolution of faults in communication networks. Integrated Network Management, III, Pages 671 – 682. Elsevier Science Publishers B.V., Amsterdam, 1993.

[LG91] P. Lucas and L. Van der Gaag. Principles of Expert Systems. Wokingham: Addison-Wesley, 1991.

[Lor93] K.-W.E. Lor. A Network Diagnostic Expert System for Acculink^{TM} Multiplexers Based on a General Network Diagnostic Scheme. H. G. Hegering and Y. Yemini (Eds.), Integrated Network Management III, North – Holland, Amsterdam, Pages 659 – 669, 1993.

[LS05] Mo Li and Kumbesan Sandrasegaran. Network Management Challenges for Next Generation Networks. Proceedings of the The IEEE Conference on Local Computer Networks 30th Anniversary, Pages 593 – 598, 2005.

[MA01] R. E. Miller and K. A. Arisha. Fault Management Using Passive Testing for Mobile IPv6 Networks. Proceedings of 2001 IEEE Global Telecommunications Conference. Vol. 3, Pages 1923 – 1927, 2001.

[Mea89] Carver Mead. Analog BLSI and Neural Systems. Addison-Wesley, 1989.

[Mi07] Nader F. Mir. Computer and Communication Networks. Pearson Hall, ISBN-13: 978-0-13-174799-9, 2007.

[MK94b] John A. Meech and Sunil Kumar. A Hypermanual on Expert Systems. Canada Centre for Mineral and Energy Technology – CANMET, Ottawa, Canada, 3rd edition, 1994.

[MK04] P. Mohapatra and S. Krishamurthy, Ad HOC Networks: Technologies and Protocols. New York, Springer, ISBN: 0387226893, 2004.

[MK05] Hermann De Meer and Christian Koppen. Self-Organization in Peer-to-Peer Systems. R. Steinmetz and K. Wehrle (Eds.), P2P Systems and Applications, LNCS 3485, Pages 247 – 266, 2005.

[MG96] Shree Murthy, J. J. Garcia-Luna-Aceves. An efficient routing protocol for wireless networks. Mobile Networks and Applications archive, Volume 1, Issue 2, Pages 183 – 197, 1996.

[MMB+02] E. Marilly, O. Martinot, S. Betge-Brezetz, and G. Delegue. Requirements for service level agreement management. Procedings of 2002 IEEE Workshop on IP Operations and Management, Pages 57 – 62, 2002.

[MS93] J. Moffett and M. Sloman. Policy Hierarchies for Distributed Systems Management. IEEE Journal on Selected Areas in Communication, Vol. 11, No. 9, Dec. 1993.

[MT99] Cristina Melchiors and Liane M. R. Tarouco. Fault Management in Computer Networks Using Case-Based Reasoning: DUMBO System. Proceedings of the Third International Conference on Case-Based Reasoning and Development, Lecture Notes In Artificial Intelligence, Vol. 1650, Pages 510 – 524, Springer-Verlag Berlin Heidelberg, 1999.

[Nej00] Wolfgang Nejdl. Metadata for Adaptive and Distributed Learning Repositories. Proceedings of World Conference on the WWW and Internet (WebNet 2000), Page 606, 2000.

[Nik00] D. Nikovski. Constructing Bayesian Networks for Medical Diagnosis from Incomplete and Partially Correct Statistics. IEEE Transactions on Knowledge and Data Engineering, Vol. 12, No. 4, Pages 509 – 516, July 2000.

[Pap09] Dimitri Papadimitriou. Future Internet: The Cross-ETP Vision Document, Version 1.0, 2009.

[Pav00] G. Pavlou. Using Distributed Object Technologies in Telecommunications Network Management, IEEE Journal of Selected Areas in Communications (JSAC), special issue on Recent Advances in Network Management and Operations, Vol. 18, No. 5, Pages 644 – 653, IEEE, May 2000.

[PB94] Charles E. Perkins and Pravin Bhagwat. Highly dynamic Destination-Sequenced Distance-Vector routing (DSDV) for mobile computers. ACM SIGCOMM Computer Communication Review, Volume 24, Issue 4, Pages 234 – 244, 1994.

[Pea88] J. Pearl. Probabilistic Reasoning in Intelligent Systems: Networks of Plausible Inference. Morgan Kaufmann, San Mateo, CA, 1988.

[Per05] Harry G. Perros. Connection-Oriented Networks: SONET/SDH, ATM, MPLS and Optical Networks, John Wiley and Sons, ISBN: 0470021632, 2005.

[PMM89] A. Patel, G. McDermott, and C. Mulvihill. Integrating network management and artificial intelligence. Integrated Network Management I, North-Holland, Amsterdam, Pages 647 – 660, 1989.

[POP01] G. I. Papadimitriou, M. S. Obaidat, and A. S. Pomportsis. Advances in optical networking. Int. J. Commun. Syst., 15:101 – 113, 2001.

[Qui93] J. Ross Quinlan. C4.5: programs for machine learning. Morgan Kaufmann Publishers, ISBN:1-55860-238-0, 1993.

[RFC3561] http://tools.ietf.org/html/rfc3561

[RFC4728] http://tools.ietf.org/html/rfc4728

[RH95] I. Rouvellou and G. W. Hart. Automatic Alarm Correlation for Fault Identification. Proceeding of IEEE INFOCOM95, Pages 553 – 561, 1995.

[Ric83] E. Rich. Users are Individuals: Individualising user models. International Journal of Man-Machine Studies. 18, Pages 199 – 214, 1983.

[RLG+06] Ridha Rejeb, Mark S. Leeson, and Roger J. Green, University of Warwick. Fault and Attack Management in All-Optical Networks, IEEE Communications Magazine, November 2006, Pages 79 – 86.

[RSR+04] Thomas Röblitz, Florian Schintke, Alexander Reinefeld, et al. Autonomic Management of Large Clusters and Their Integration into the Grid. Journal of Grid Computing (2004) 2: 247 – 260, Springer, 2004.

[Sch03] Albrecht Schmidt. Ubiquitous Computing – Computing in Context. PhD dissertation, Lancaster University, 2003.

[SCH+04] N. Sebe, I. Cohen, T. S. Huang, and T. Gevers. Skin Detection: A Bayesian Network Approach. Proceedings of the 17th International Conference on Pattern Recognition (ICPR 2004). Volume 2, Pages 903 – 906, 2004.

[Sch07] Jürgen Schönwälder, Internet Management Protocols. Handbook of network and system administration, Jan Bergstra and Mark Burgess (Eds.), ISBN-13: 978-0-444-52198-9, ELSEVIER, 2007.

[SDR04] Andreas Schmietendorf, Reiner Dumke, and Daniel Reitz. SLA Management – Challenges in the Context of Web-service-based Infrastructures. Proceedings of the IEEE International Conference on Web Services, Pages 606 – 613, 2004.

[SFD+09] J. Schöwälder, M. Fouquet, G. Dreo Rodosek, and I. Hochstatter. Future Internet = Content + Services + Management. IEEE Communications Magazine, Pages 27 – 33, July, 2009.

[She96] K. Sheers. HP OpenView Event Correlation Services. Hewlett-Packard Journal. Vol. 47, No. 5, Pages 31 – 42, Oct. 1996.

[SJB00] A. Singhal, Luo Jiebo, and C. Brown. A multilevel Bayesian Network Approach to Image Sensor Fusion. Proceedings of the Third International Conference on Information Fusion. Vol. 2, Pages 9 – 16, 2000.

[Sla91] S. Slade. Case-base reasoning: A research paradigm. AI magazine, 12(1): 42 – 55, 1991.

[SMA01] Automating Root Cause Analysis: Codebook Correlation Technology vs. Rule Based Analysis, White Paper, SMARTS, http://www.smarts.com, 2001.

[SS02] M. Steinder and A. S. Sethi. Non-deterministic diagnosis of end-to-end service failures in a multi-layer communication system. In Proc. of ICCCN, Pages 374 – 379, 2001.

[Str03] John S. Strassner. Policy-based Network Management: Solutions for the Next Generation, ISBN: 1558608591, 9781558608597, Morgan Kaufmann, 2003.

[Sun05] Zhili Sun. Satellite Networking Principles and Protocols. ISBN-13 978-0-470-87027-3, John Wiley & Sons, 2005.

[TDM05] Ryan W. Thomas, Luiz A. DaSilva, and Allen B. MacKenzie. Cognitive Networks. Proceedings of First IEEE International Symposium on New Frontiers in Dynamic Spectrum Access Networks, Pages 352 – 360, 2005.

[TEM04] Ulf Troppens, Rainer Erkens,and Wolfgang Müller. Storage Networks Explained: Basics and Application of Fibre Channel SAN, NAS, iSCSI and InfiniBand, ISBN: 0470861827, Wiley, 2004.

[TMK06] F. Travostino, J. Mambretti, and Gigi Karmous-Edwards. Grid Networks: Enabling Grids with Advanced Communication Technology. ISBN-13: 978-0-470-01748-7, John Wiley & Sons, 2006.

[Toh97] Chai-Keong Toh. Associativity-Based Routing for Ad Hoc Mobile Networks, Wireless Personal Communications, Vol. 4, No. 2., Pages 103 – 139, 1997.

[TTA08] K. Tutschkua, P. Tran-Gia, and F.-U. Andersen. Trends in network and service operation for the emerging future Internet. Int. J. Electron. Commun. (AEÜ) 62 (2008), Pages 705 – 714.

[Ver02] D. C. Verma. Simplifying network administration using policy-based management. Network, IEEE, Vol. 16, Pages 20, 2002.

[WBE+98] P. Wu, R. Bhatnagar, L. Epshtein, M. Bhandaru, and Z. Shi. Alarm correlation engine (ACE). Proceedings of Network Operation and Management Symposium (NOMS'98), Pages 733 – 742, 1998.

[WEW99] G. Weiss, J. Eddy, and S. Weiss. Intelligent Telecommunication Technologies. In Knowledge-based Intelligent Techniques in Industry. L. Jain, R. Johnson, Y. Takefuji, and L. Zadeh (Eds.), Pages 251 – 275, CRC Press, 1999.

[WHB05] Weiss, E., Hiertz, G.R. and Bangnan Xu. Performance analysis of temporally ordered routing algorithm based on IEEE 802.11a. Proc. of IEEE 61st Vehicular Technology Conference, Volume: 4, Pages 2565 – 2569, 2005.

[Wie02] H. Wietgrefe. Investigation and practical assessment of alarm correlation methods for the use in GSM access networks. R. Stadler and M. Ulema (Eds.), Proceedings of Network Operation and Management Symposium (NOMS'02), Pages 391 – 404, April 2002.

[WKT+99] W. Wiegerinck, H. J. Kappen, E. W. M. T ter Braak, W. J. P. P ter Burg, M. J. Nijman, Y. L. O, and J. P. Neijt. Approximate Inference for Medical Diagnosis. Pattern Recognition Letters, 20: 1231 – 1239, 1999.

[WS93a] C. Wang and M. Schwartz. Fault detection with multiple observers. IEEE/ACM Transactions on Networking, Vol. 1, Pages 48 – 55, January 1993.

[WS93b] C. Wang and M. Schwartz. Identification of faulty links in dynamic-routed networks. Journal on Selected Areas in Communications, 11(3): 1449 – 1460, Dec. 1993.

[WTM95] Andrew J. Weiner, David A. Thurman, and Christine M. Mitchell. Applying Case-based Reasoning to Aid Fault Management in Supervisory Control. Proceedings of the 1995 IEEE International Conference on Systems, Man and Cybernetics, Pages 4213 – 4218, 1995.

[YKM+96] S. A. Yemini, S. Kliger, E. Mozes, Y. Yemini, and D. Ohsie. High speed and robust event correlation. IEEE Communications, 34(5): 82 – 90, May 1996.

[Zak] Robert H'obbes' Zakon. http://www.zakon.org/robert/internet/timeline/

[ZWX+04] Jihui Zhang, Jinpeng Huai, Renyi Xiao, and Bo Li. Resource management in the next-generation DS-CDMA cellular networks. IEEE Wireless Communications, Volume 11, Issue 4, Pages 52 – 58, 2004.

Index